Excel

YEAR 9

Science Revision Workbook

ESSENTIAL skills

Get the Results You Want!

PASCAL PRESS

Donna Bennett

Reprinted 2002, 2003, 2004, 2006, 2008, 2009, 2011

Updated in 2015 for the Australian Curriculum

Reprinted 2019, 2022, 2025

ISBN 978 1 74020 080 6

Pascal Press
PO Box 250
Glebe NSW 2037
(02) 9198 1748
www.pascalpress.com.au

Publisher: Vivienne Joannou
Project editor: Mark Dixon
Edited by Shelley Barons and Alicia Thompson
Australian Curriculum updates edited by Geoffrey Thickett
Typeset by Scriptorium Desktop Publishing Pty Ltd and Kim Webber
Cover and page design by Dizign Pty Ltd
Printed by Vivar Printing/Green Giant Press

Contents

Introduction

This book is suitable for students studying the Year 9 Australian Curriculum Science course. It is particularly valuable for those students who are interested and motivated to seek an excellent result in Science.

This is a user-friendly revision workbook that students may use throughout the year to support classwork and help them with their revision for tests and exams.

The features of this book are as follows:

- The eight topics conform to the Australian Curriculum Science syllabus.
- Each topic consists of a concise summary of the key concepts in the topic.
- Each topic summary is followed by several pages of skill and knowledge practice questions to further enhance understanding.
- There are two sample Test Papers which bridge four topics and are aimed at the consolidation of knowledge.
- Answers to all questions are provided together with a complete glossary.
- Questions provide valuable revision before a test or exam.
- The book includes questions to interest the curious, capable student.

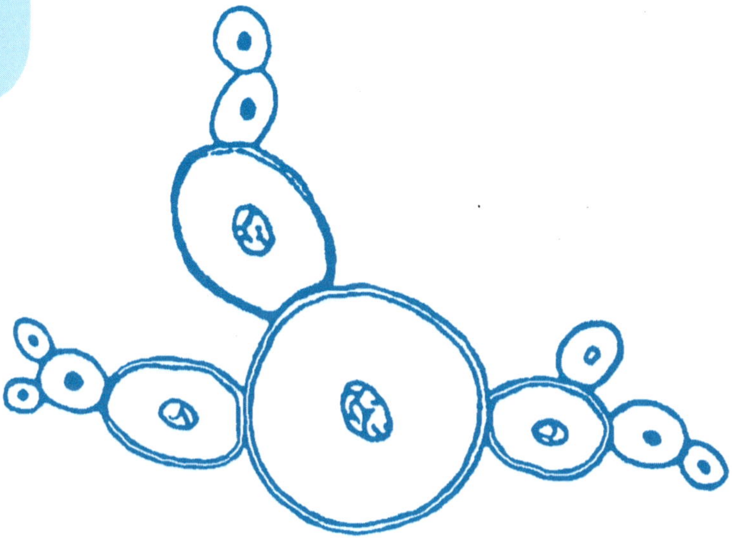

CHAPTER 1

FOOD AND LIVING THINGS

On which group of living things does all life depend?

Some living things can make their own food and others must consume their food.

Plants

Plants provide the foundation for life on Earth. Plants are unique among living things as they can make their own food. Without plants, many other living things, including humans, could not exist. Plants also provide wood, paper, fibres, habitats and natural beauty.

Photosynthesis

- The leaves of plants are like small solar panels that gather the sun's energy and use it for a food-making process called **photosynthesis**.

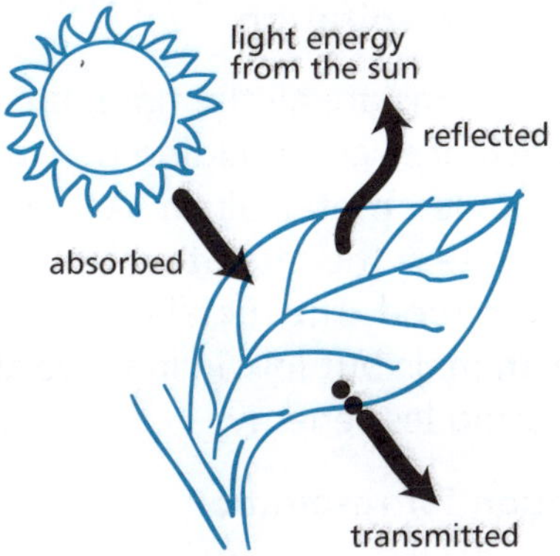

- The food made by photosynthesis are simple sugars like glucose and fructose. These sugars can be stored as starch and then broken down to provide the energy used to build up leaves, stems, roots and flowers. Photosynthesis takes place in the chloroplasts inside plant cells. During photosynthesis, a plant uses the energy in sunlight to carry out a series of chemical reactions. It makes the food, a substance called glucose, from molecules of carbon dioxide and water. Oxygen is formed as a waste product.

Word equation for photosynthesis

carbon dioxide + water → glucose + oxygen

A single molecule of glucose is formed by combining six water molecules with six carbon dioxide molecules, using energy from the sun. Six oxygen molecules are also formed.

Chemical equation for photosynthesis

$6CO_2 + 6H_2O \rightarrow C_6H_{12}O_6 + 6O_2$

- A plant traps the light it needs with special chemicals called **photosynthetic pigments**. The main pigment in green plants is chlorophyll. **Chlorophyll** gives plants their green colour because it reflects green light and absorbs red and blue light.

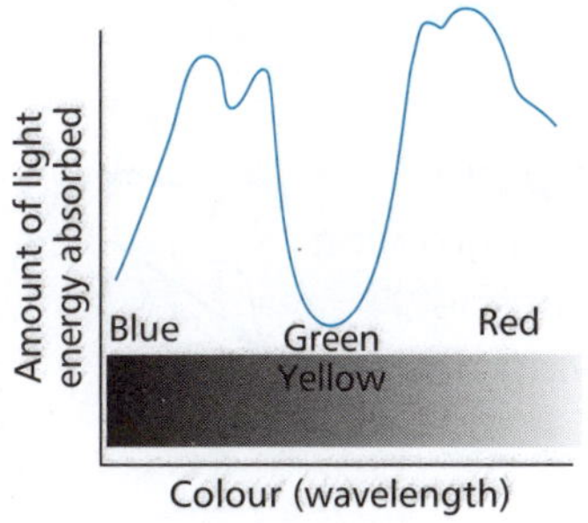

Chlorophyll absorption pigment

Plant growth

Unlike animals, plants cannot move around to find food but they can reach towards light and minerals in the soil, and react to the environment around them.

- Living things grow mainly by increasing their number of cells through cell division. Plants can also grow by making existing cells bigger. Some plants stop growing once they have reached a certain size but others continue to grow as long as conditions are good.
- Some cells stop dividing once they are mature but **meristematic cells** continue to divide throughout a plant's life. These cells can be found at the tip of a root or stem and are responsible for the increase in length of a stem or root.

The growth of a plant

Seedling's cells divide. The seed provides the food.

Cells divide at the seedling's meristems.

Once a plant part is formed it stops growing but the cells at the plant's growing tips still divide.

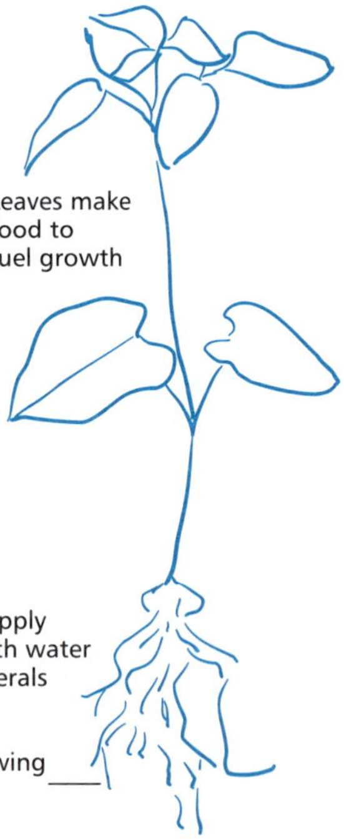

Animals

Animals also need energy for growing, moving and responding to the changing environment. **Oxidation** is the process of reacting substances with oxygen to produce energy. Most animals combine food with oxygen in the same way during a process that biologists call **respiration**.

Respiration

- Energy is stored in chemical substances like food and these must be broken down to release the energy. Respiration is a series of chemical reactions that take place inside the cells of animals. During respiration, organic substances like glucose are broken down to release energy. Plants respire at night to get energy when there is no sunlight for photosynthesis.

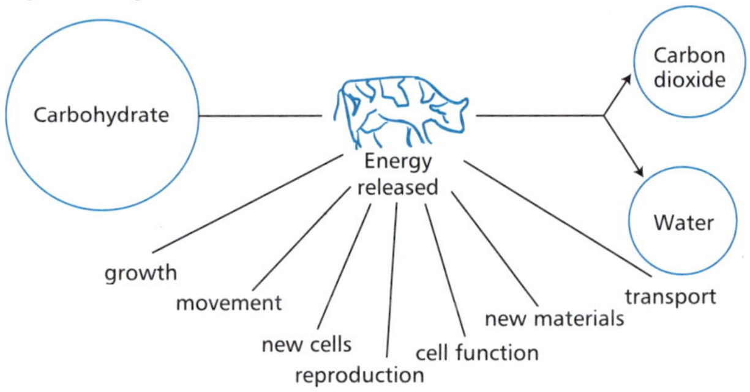

The process of respiration

- Living organisms are all the time, breaking down substances and building new ones. This process is called **metabolism**. An animal's metabolic rate is the rate at which it releases energy from food. The metabolic rate is high in active animals but low in inactive animals that are using less energy.

Word equation for respiration

glucose + oxygen → water + carbon dioxide + energy

Chemical equation for respiration

$$C_6H_{12}O_6 + 6O_2 \rightarrow 6CO_2 + 6H_2O + \text{energy}$$

- The word 'respiration' is also used to describe the process of getting oxygen into an animal's body. This process involves organs like lungs in mammals, gills in fish and spiracles in insects.

Higher organisms have special systems of organs called respiratory systems for getting the oxygen from the outside environment to individual cells.

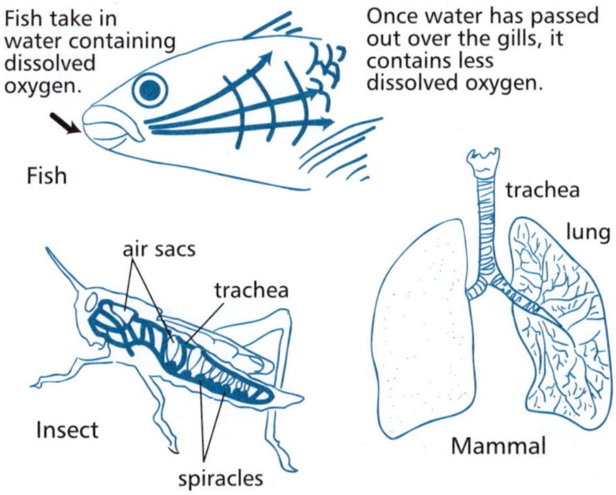

Respiratory organs in mammals, fish and insects

Feeding

- Animals have to take in food from their surroundings to survive. They do this by feeding or eating. This process is called **ingestion**.
- Once an animal has ingested some food the food is broken down by a process called **digestion**. Multi-celled organisms have special organs that digest the food. Humans have teeth, a stomach and intestines to help break the food down.

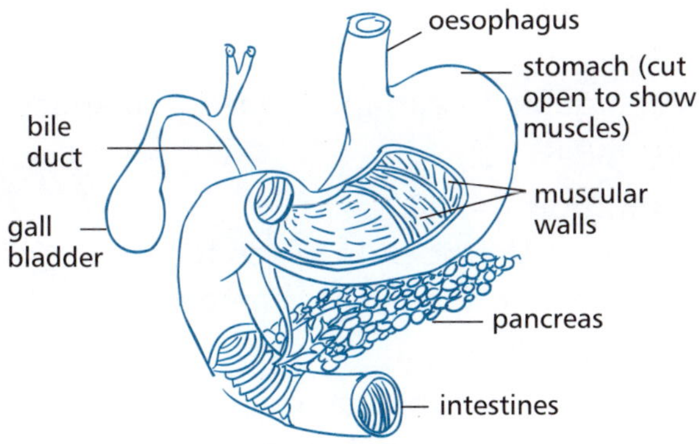

Digestive system of a human

Nutrition

To stay healthy humans need to eat the right sorts of foods in the right amounts. This is called a balanced diet. People from different countries get healthy diets from all sorts of foods. It is generally agreed that salt, sugar and fat should be consumed in small amounts.

A healthy diet must contain:

- **Carbohydrates**, which supply energy. Foods such as whole grains, pasta, potatoes and cereals are sources of carbohydrates. Carbohydrates are made of hydrogen, oxygen and carbon grouped together to form molecules. The smallest parts of carbohydrates are **sugars**. Energy is measured in joules (J) or kilojoules (kJ).
  1000 joules make one kilojoule.

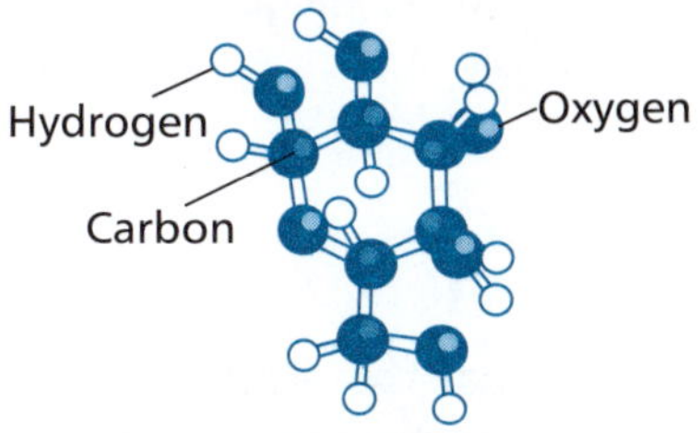

Carbohydrate molecule

- **Fibre** is found in fruit and vegetables. This is that part of the plant that cannot be digested by humans but it helps prevent disease.
- **Proteins** are essential for growth and to repair cells and tissues. Left over protein is used for energy. Foods that are high in protein are meat, fish, eggs, milk and cheese. Proteins are groups of hydrogen, oxygen, carbon and nitrogen. The smallest parts of proteins are **amino acids**.

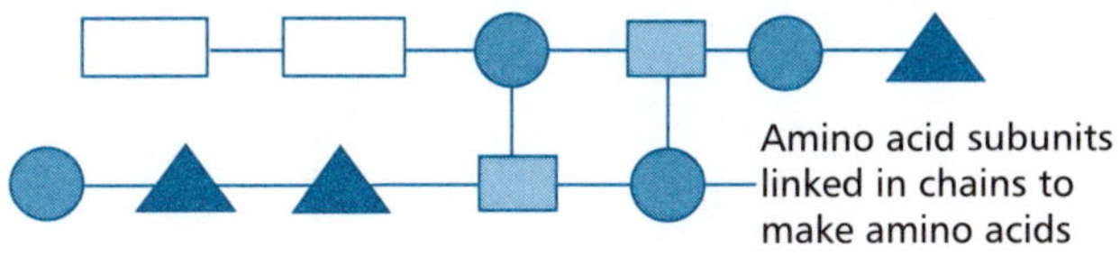

Structure of proteins

- **Fats and oils** are lipids that provide energy, insulation and fat-soluble vitamins. Lipids are composed of chains of hydrogen, oxygen and carbon, arranged in molecules called triglycerides. A triglyceride is made of three fatty acids combined with glycerol. Fats release twice as much energy as carbohydrates.

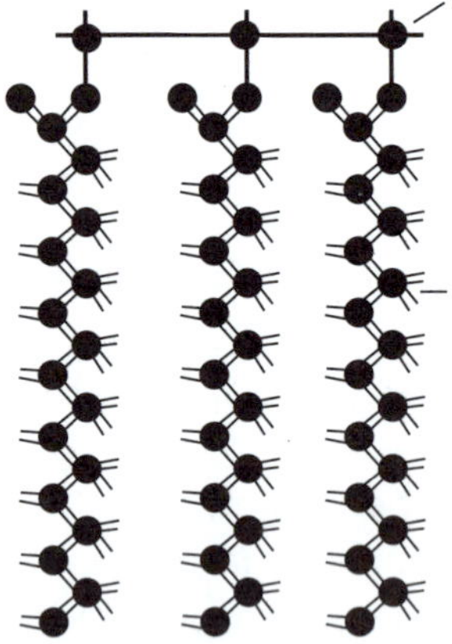

Triglyceride molecule

- **Vitamins and minerals** help the body metabolise foods.
- **Water**

Healthy eating

It is important to eat the right foods in the right amounts to prevent disease and illness.

- Too much food can cause people to be fat which puts extra strain on the heart.
- Too much sugar may cause bad teeth.
- Not enough iron can cause anaemia.
- Not enough vitamin A could cause poor eyesight.
- Too little calcium may cause weak bones.

Nutrition labelling on packaging in Australia can help the consumer compare the nutritional value of foods when purchasing processed foods.

Glossary

Carbohydrates
An organic compound that contains hydrogen, oxygen and carbon.

Chlorophyll
The main photosynthetic pigment of plants.

Digestion
The process whereby food is broken down into its smallest parts.

Fibre
An insoluble carbohydrate that plants use as a building material.

Ingestion
Taking food from the surrounding environment into the body.

Metabolism
All the chemical processes that take place in a living thing.

Oxidation
Oxygen joining with other elements to make other substances, such as in burning, respiration and rusting.

Photosynthesis
A process that uses light energy to make food from simple chemicals.

Photosynthetic pigments
A chemical that collects the energy of sunlight.

Protein
An organic compound made up of amino acid units.

Respiration
A chemical process in which food is broken down to release energy.

Respiratory system
A group of organs that transports oxygen from the exterior to the cells of the body.

Sugars
Simple carbohydrates that have a sweet taste.

Science skills

Question 1

Label the diagram below with the following phrases:

A Carbon dioxide molecule is absorbed from the air into the leaves.

B Water molecules are drawn up from the roots.

C Energy from the sun is the force behind photosynthesis

D Oxygen molecules are a by-product of photosynthesis.

E Glucose molecules are produced by photosynthesis.

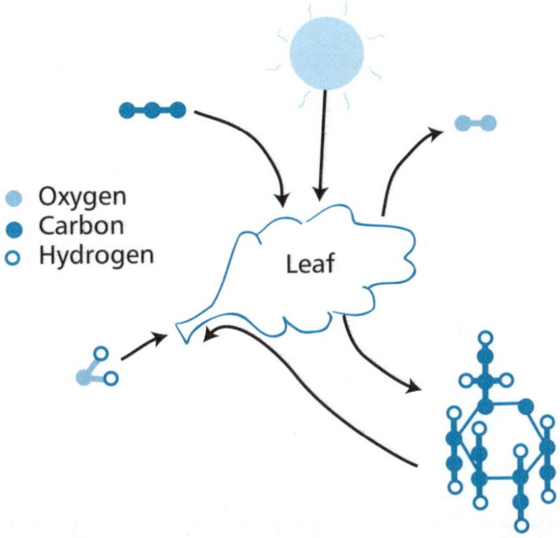

Question 2

A new plant will grow from a seed.

Arrange the phrases below to show the life cycle of a plant:

The seedling grows using the food stored in the seed.
A seed germinates.
The flowers make seeds.
The plant flowers.

Question 3

Decide which meals are healthy and mark the appropriate column.

Meal	Healthy	Unhealthy
pizza		
chicken and vegetables		
fish fingers and chips		
steak and salad		
stir fry vegetables		

Question 4

Match the following foods and food groups.

Carbohydrates	vegetables, fruit
Protein	olive oil, dairy products
Lipids	cereals, potatoes, pasta
Fibre	meat, eggs, fruit

Question 5

Study the table below:

Milk Protein and Growth

Animal	Protein %	Time taken to double birth weight
human	1.4	180 days
cow	3.5	45 days
dog	7.4	9 days
cat	7.0	9.5 days
rabbit	10.4	6 days

a Write a statement about the relationship that appears to exist between growth rate and protein in the mother's milk.

b Draw a graph, plotting the time taken to double the birth weight and % protein.

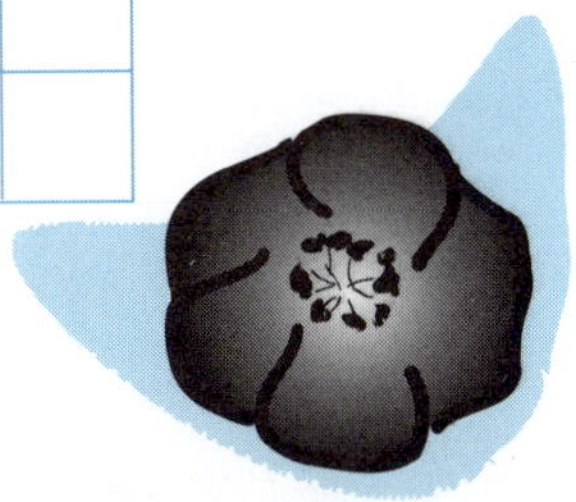

Question 6

The following table lists the energy, protein and fat levels of some common drinks.

Drink	Energy (kJ)	Protein (g/serve)	Fat (g/serve)
Sprite	383	–	–
Coca-cola	393	–	–
chocolate shake	1508	11.9	9.5
vanilla shake	1446	11.2	9.1
orange juice	302	1.1	0.7
white tea	51	0.7	0.7
white coffee	106	1.4	1.4

a Which drink provides the most energy?

b Which drink contains the least protein?

c How much fat does orange juice contain?

Question 7

Study the following graphs and then answer the questions.

Daily energy requirements of men and women

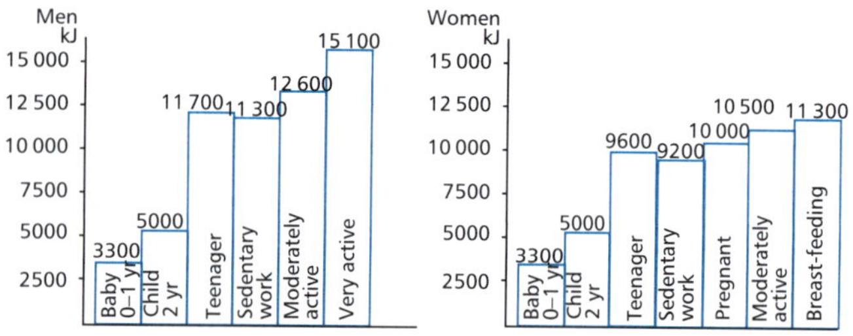

What is the daily energy requirement for a:

a pregnant woman?

b teenage boy?

c moderately active woman?

d very active man?

Question 8

By reading the table below two students (A & B) calculated the total kilojoules in their lunches.

Food	Quantity	Energy (kJ)
sausage roll	1	1987
white bread	2 slices	470
tomato	1	101
butter	teaspoon	290
doughnut	1	731
fruit cake	1 slice	508
malted milk	glass	1554
orange juice	glass	479

Lunch A — 2 sausage rolls
1 doughnut
1 malted milk

Lunch (B) — 2 tomato sandwiches
1 piece of fruitcake
1 orange juice

a How many kilojoules in lunch A?

b How many kilojoules in lunch B?

c Which student consumed the most kilojoules?

Question 9

Using the data from question 8 and the information in the table below, answer the questions.

Activity	Energy used in kilojoules per minute
walking	20
squash	59

a For how long would you have to walk to use up the energy from lunch B?

b How much faster would you use up this energy if you were playing squash?

Question 10

Label the packet of Oaties with the information that must be on the pack.

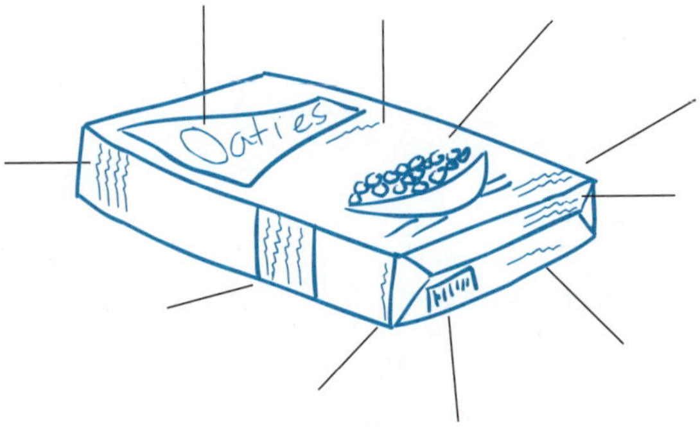

Information on the pack:

a Trade name

b Name that exactly describes the food

c Accurate picture

d Net weight (minus the packaging)

e Name and address of the manufacturer

f Use-by date

g Bar code

h Storage instructions

i Preparation instructions

j Nutrition claim and all ingredients

Question 11

The following table lists some of the vitamins (micrograms) and minerals (milligrams) in foods.

Food	Vitamin A	Vitamin B	Calcium	Iron
milk	0.8	0.4	12.0	0
beef	0	0.7	1.1	0.4
eggs	6.0	1.5	5.6	0.2
rice	0	0.7	0.4	0.1
fish	0.8	0.4	10.2	0.1
bread	0	2.1	2.5	0.3

a Name one good source of vitamin A.

b What condition develops if vitamin A is deficient in the diet?

c Name two good sources of vitamin B.

d Which food has the highest level of iron?

e A lack of which mineral causes anaemia to develop?

f Which foods are high in calcium?

g If a diet is low in calcium what part of the body may become weaker?

Question 12

Complete the table below by marking the correct food category.

Food	Carbohydrate	Protein	Fat
milk			
carrots			
cereal			
eggs			
butter			
apples			
fish			
pasta			
chicken			

Question 13

Which of the following foods could be considered to be 'healthy'?

Question 14

Mark on the drawing below the position of rapidly dividing cells.

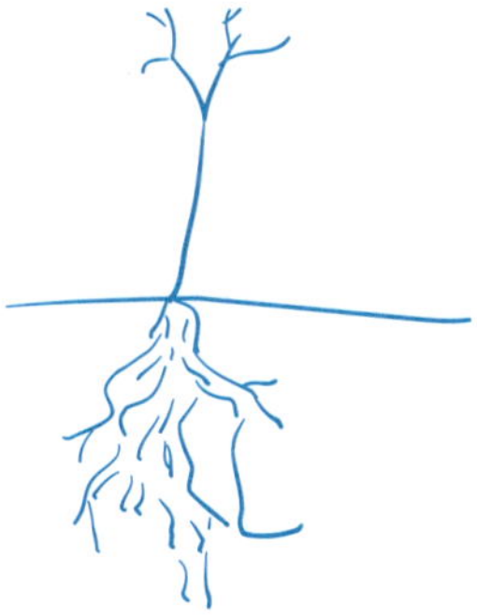

Question 15

Study the graphs below and answer the questions.

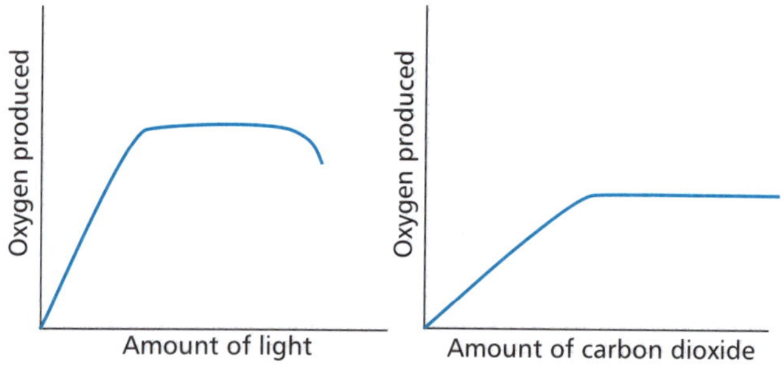

a What happens to the rate of photosynthesis when the amount of light decreases?

____

____

____

b What happens when the carbon dioxide levels increase?

____

____

____

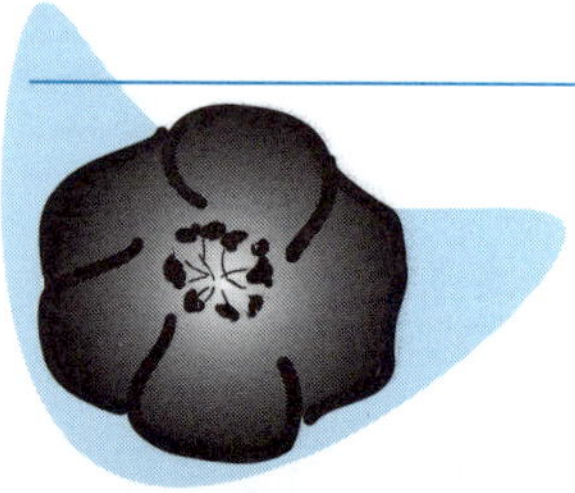

Revision questions

Question 1

What is a balanced diet?

____

____

Question 2

Fill in the missing words.

Your __________ system is a group of __________ that breaks down __________ into smaller portions so the body can use it.

Question 3

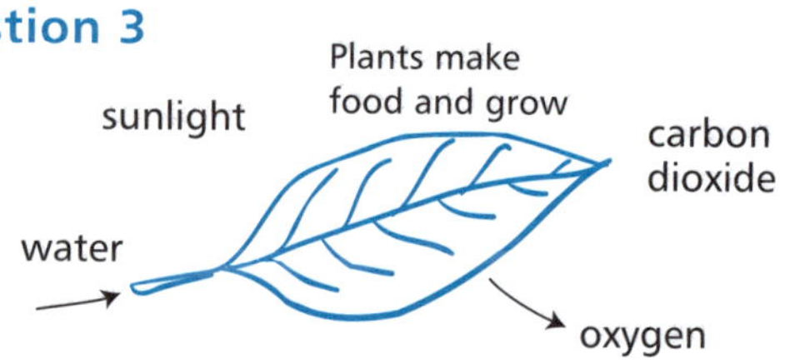

Chlorophyll in the ______________ of leaf cells use energy from sunlight, __________ __________ and __________ to make food. This process is called ______________.

Roots take in water and ______________ from the soil.

Question 4

a Explain how a dingo ultimately depends on green plants for its food?

____

____

____

b Use a dictionary to find out what the terms *photo* and *synthesis* mean.

____

____

c Glucose is one of the products of photosynthesis, what is the other?

Question 5

a Write the chemical equation for photosynthesis.

b Write the chemical equation for respiration.

c Compare these equations — how are they similar?

Question 6

Plants need minerals for growth.

Study the table below and work out which minerals are important for photosynthesis.

Mineral	What it does	If there is not enough
nitrogen	main part of chlorophyll, important in cell growth	yellow, stunted plants with a small leaf area
phosphorus	for germinating seeds, root and fruit development	poor roots, stunted growth, delayed seed development
potassium	important for food production and disease prevention	small seeds and fruit, weak stems, poor leaves

Question 7

What are the basic building blocks for:

protein — ______

carbohydrates — ______

fats — ______

Question 8

Foods are sometimes divided into the following groups. For a healthy diet write *eat moderately*, *eat lots*, or *eat small amounts* against the following food groups.

a bread, cereals, fruit and vegetables

b milk, cheese, meats, nuts, yoghurt

c butter, cream, salt, sugar

Question 9

Which of the following foods do you think are high in fibre?

muesli, sponge cake, sultanas, wholegrain bread, baked beans, instant potato, bananas

Question 10

a What elements are in lipids?

b Are there any elements in lipids that are not in carbohydrates?

c What element is in protein that is not in carbohydrates?

Question 11

a Why would a plant grow through a maze like the one below just to get to the light?

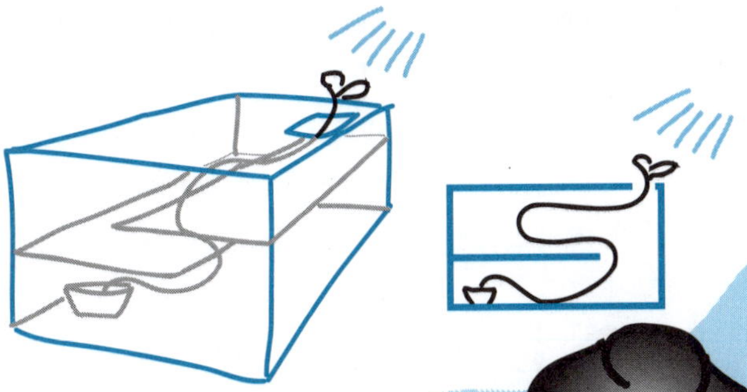

b Finish drawing this plant. Add more stem and 6 more leaves.

Question 12

Many classrooms have fish tanks like this one.

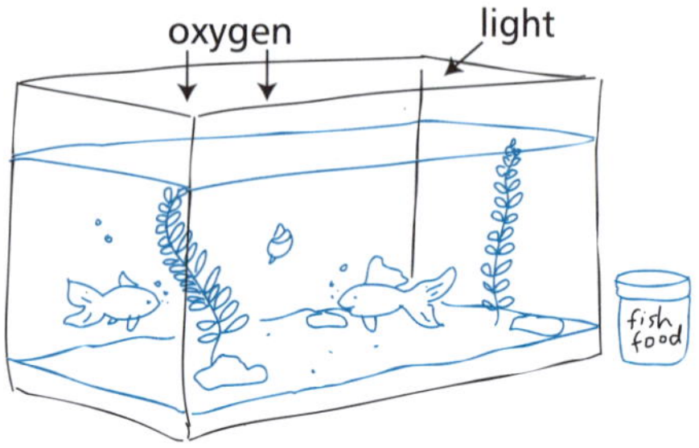

a Complete the table to show what the fish and the plants need to stay healthy.

To stay healthy need	Water weed	Fish	Snail
oxygen			
carbon dioxide			
sunlight			
water weed			
fish food			

b Sometimes fish tanks have a pump that bubbles air into the water. What essential element is this adding to the water?

____

Question 13

The following is a list of energy changes. Put them in the correct order.

_______ Energy in an animal's body

_______ Energy in glucose

_______ Energy in sunlight

_______ Energy in starch

_______ Energy in movement

Question 14

A group of students wanted to get the energy out of a peanut. They speared the peanut with a mounted needle, set it alight and used the flame to heat the water in a test tube.

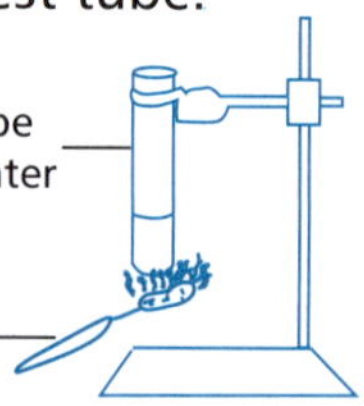

Put the following phrases in the correct order and join them with arrows to show the movement of heat energy into the water. Then put the words *photosynthesis* and *oxidation* beside the arrows that represent these processes.

Phrases:
sunlight energy
heat energy in the water
energy in glucose in the peanut leaf
energy in the oil in the peanut

Question 15

Respiration is a chemical reaction.

a Where does it take place?

____

b What type of chemical reaction is it?

____

c Why is it so important to living cells?

____

____

d Respiration releases energy from food. Explain how the energy came to be in the food in the first place.

____

____

____

Question 16

Every animal cell needs oxygen.

a Which main organs are responsible for getting oxygen into the human body?

____

b What is the name of this system?

Question 17

a Why do we need carbohydrates in our diet?

b Name two foods very high in fat.

c Why should everyone have at least some fat in their diet?

d Which types of foods provide the most energy?

e What happens if the amount of energy taken into a person's body is more than the amount of energy they use up?

Question 18

A fifteen-year-old boy ate the following foods in one day.

Breakfast — nothing
Morning tea — potato chips and soft drink
Lunch — hamburger, chips, soft drink, chocolate
Dinner — fish and chips, piece of cake, 3 chocolate biscuits, ice cream

a Is this a balanced diet? ____________

b What could he do to improve this diet without giving up the things he likes?

Question 19

a Complete the table by adding three examples for each food group.

Food group	Examples
carbohydrates	
fatty foods	
proteins	
fibrous foods	

b You need the right amount and types of food to stay ____________. You also need to drink plenty of __________. The other important things in you diet are c____________, f__________, p__________, f__________, v____________, m__________.

Question 20

Below is a list of the major minerals that some crop plants need to grow (kilograms per hectare). Study the data and answer the questions.

Crop	Nitrogen	Potassium	Phosphorus
wheat	85	20	20
peanuts	110	35	9
cotton	135	45	32
potatoes	140	225	24
lucerne	710	580	67

a Which crop has the highest demand for nitrogen per hectare?

b Which mineral shows the greatest variation in demand?

c When a crop is harvested the minerals are removed. What is the total weight of minerals removed with one hectare of peanuts?

d How much potassium does a hectare of wheat need?

e How much phosphorus does a hectare of potatoes need?

CHAPTER 2

SENSING THE WORLD AROUND YOU

Does the nose know?

All animals rely on their **senses** to find out about what is going on in the world around them so they can act or react appropriately. For example our sense organs tell us that fire is hot and that a knife is sharp.

'Reality' is constructed in our brains according to the information that is available. Some creatures depend on only one or two senses, for example fish and birds have very well developed eyesight.

In humans most senses are developed and they tend to complement each other. We generally identify five basis human senses — sight, hearing, touch, smell and taste.

The five senses

Vision

Vision, or **sight**, is very important because we can see things that are near and far. Touch, taste and smell are useful only for things that are close to us.

Vision is the sense that can detect light. Most animals can tell the difference between light and dark. Animals with more advanced vision can form an image of the environment around them. This image is recorded by the eye upside down, but the brain turns it up the right way.

- **Structure of the eye**

  The eye is the organ that detects the light reflected from or given off by an object. Eyes have special cells called **photoreceptors** that respond to light by sending signals to the brain via the **optic nerve** when light falls on them. In vertebrates these photoreceptors are packed together in a membrane called the **retina**. The signals include information about the brightness, colour and size of an object. The retina of mammals has an area called the blind spot where the optic nerve leaves the retina.

  All light entering the eye passes through the small black **pupil** in the centre. The coloured **iris** around the pupil automatically adjusts to limit or increase the amount light that enters the eye. The iris contracts in bright light and expands in dim light.

  The **cornea** is a thin, clear disc that protects the pupil. The **lens** is just inside the pupil. Muscles attached to the lens change the curvature of the lens to focus the image onto the retina.

  Most of the eye contains a clear, jelly-like substance called the **vitreous humour** and between the lens and the cornea is a more watery fluid called the **aqueous humour**.

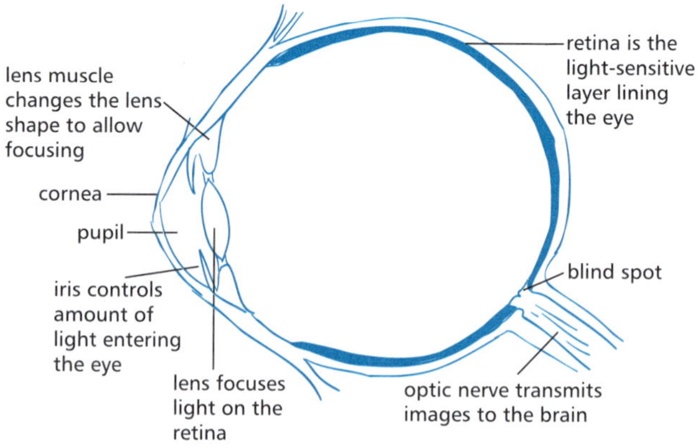

Structure of the eye

The entire eyeball is protected by a bony socket called the orbit, and the eyeball is washed by tears. Tears lubricate the eye and help to kill bacteria.

- **Defects of vision**

  This is a characteristic of the eye that prevents it forming a sharp image. Normally the lens focuses images directly on your retina. Nearsighted people can't focus on objects far away because their lens focuses images in front of the retina. Glasses with a concave (curved inwards) lens correct this problem. Farsighted people can't focus on objects close to them because their lens focuses behind their retina. Glasses with a convex (curved outwards) lens correct this problem.

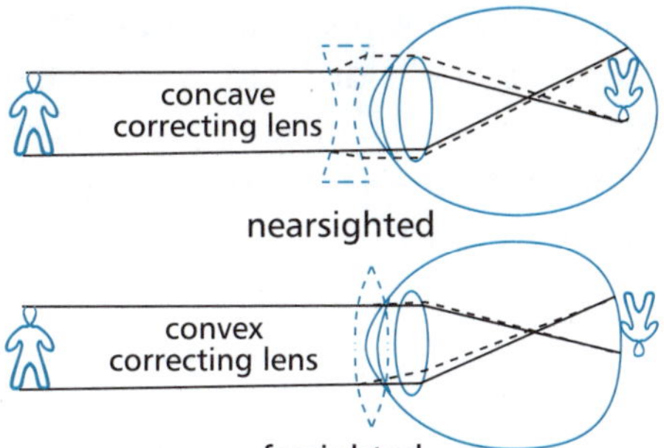

Correcting lenses

Hearing

Perhaps the most important feature of hearing is that it allows animals to communicate and makes human speech possible.

The sounds you hear are vibrations of the air around you; the louder the noise, the more the air vibrates.

- **How does the ear work?**

  The outer part of your ear, called the pinna, collects and channels the vibrations down the auditory canal to a thin membrane that is stretched very tightly called the **eardrum**.

  The middle ear is an air filled space in the skull that contains the ear bones. The eardrum vibrates and the vibrations are amplified by three tiny bones — the hammer, the anvil and the stirrup. These bones transfer the vibrations to the inner ear which is deep inside your head.

  The inner ear consists of a coiled tube called the **cochlea** and the organs of balance. The cochlea is filled with fluid and divided down its length by a flexible membrane. Along the membrane are thousands of cells which have stiff hairs sticking up from their surface. These hair cells transmit the vibrations to the brain which registers them as sound.

  Just as we need two eyes to work out where objects are in relation to each other, we need two ears to locate sound.

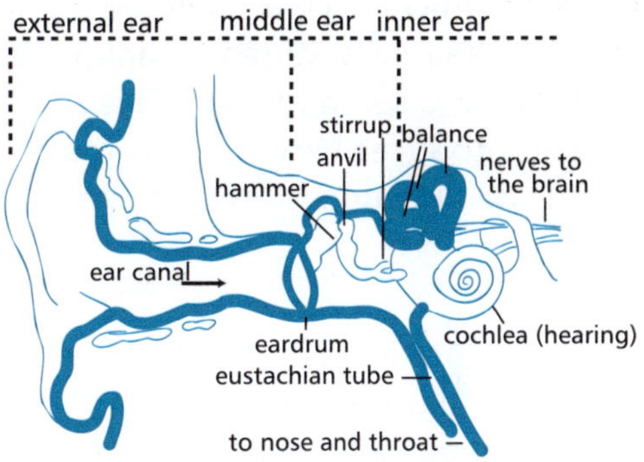

Structure of the ear

- **Echolocation**

  Animals that use high frequency sound to build up a picture of their surroundings include bats, dolphins and some whales. They send out high-pitched squeaks that bounce off anything in their path and return to the animal as echoes. The animal's brain then analyses the echo to form an image and to estimate how far away the object is.

- **Damage to hearing**

  Hearing impairment can be caused by many different things. Examples are blockage (wax) or damage to the external or middle ear, infection of the ear canal, perforated eardrum, damage to the ear bones or hair cells as well as hereditary hearing impairment. Sometimes people with damaged cochleas are given a cochlea implant (bionic ear) to improve their hearing.

Touch and balance

Touch and balance tell an animal about its immediate surroundings and about the movement and position of the body.

Touch works mainly through the skin — it is your biggest sense organ. It has millions of nerve endings, which are sensitive to light, heavy pressure, heat, cold and pain. Some parts of the body, like hands, are packed with nerve endings, making them more sensitive than other parts.

Balance is a sense that detects gravity and movement. The organs of balance are located in the inner ear. In mammals, each ear has three **semicircular canals** which are filled with fluid and lined with hair cells. When an animal moves its head the fluid is set in motion. This bends the hair cells and produces nerve signals that tell the brain about the direction of movement.

Taste

Taste and smell are the senses that detect chemicals. They allow animals to communicate, find food and avoid substances that are harmful.

When an animal tastes something, cells called chemoreceptors respond to chemicals dissolved in saliva.

A taste bud is a small sense organ which contains a cluster of chemoreceptors connected to the brain via nerves. The human tongue has about 2000 taste buds that are sensitive to sweet, sour, salt and bitterness.

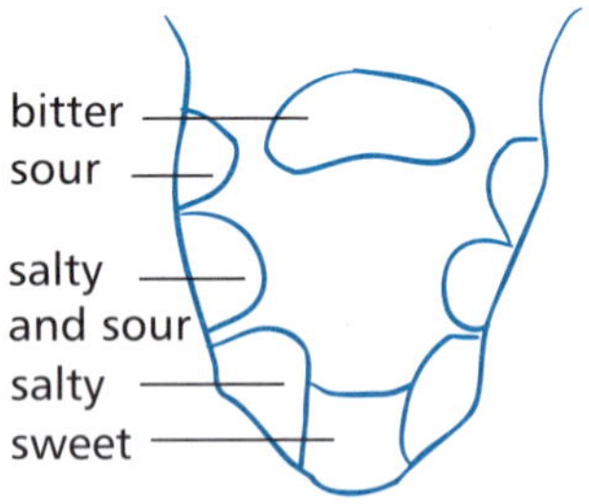

A map of the tongue

Smell

Smell is a sense that detects airborne chemicals.

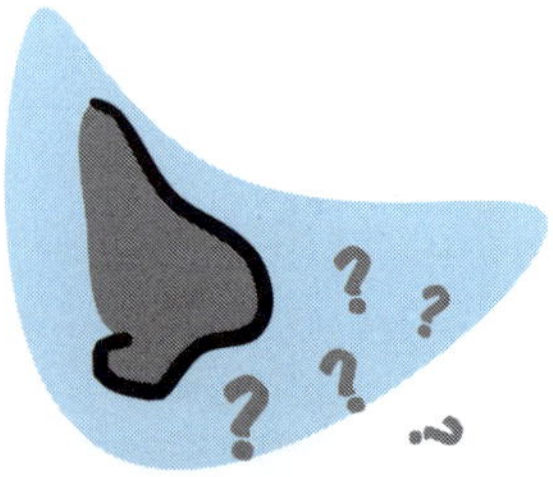

Land animals use their sense of smell, or **olfactory sense**, to detect these chemicals.

- The human sense of smell is much more highly developed than taste. But our noses are not nearly as sensitive as a dog's nose.
- Smell and taste work together to identify the taste of food. If you have a cold the flavour of food is reduced as the sense of smell is not working as well.
- The olfactory membrane is a thin moist membrane covered in densely packed chemoreceptors that end in olfactory hairs. The hairs detect different chemicals in the air and send signals to the brain.

Other sense monitors

- Jacobson's organ is a special organ that snakes use to detect smells or taste. Snakes flick their tongues in and out, collecting chemicals that they transfer to the Jacobson's organ in the roof of their mouth.

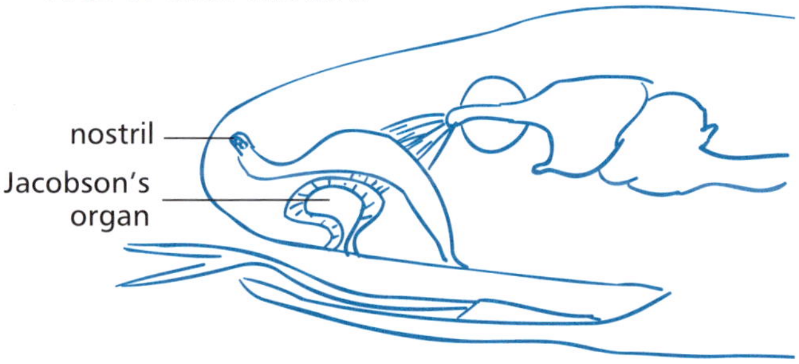

Snake — Jacobson's organ

- Antennae are sensory structures on the heads of arthropods. They can detect air movements, vibrations and air borne chemicals. Male moths often have antennae that can detect chemicals called pheromones which are important in attracting a mate.

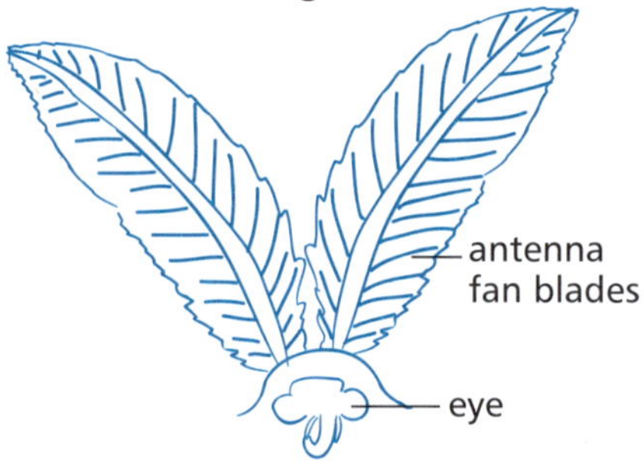

The antennae of a moth

- The senses of some animals are adapted to the time of day that they hunt. For example a cat has specialised eyes to gather what little light there is at night for hunting.

How do the senses function?

The nervous system allows an animal to respond quickly to a stimulus from its environment.

- A receptor is a specialised nerve cell that converts physical stimulation into electrical information that the brain can use. Receptors are scattered throughout the body or clustered in groups in the sense organs.

  Receptors can be stimulated in only two ways: chemically as in taste, smell and vision; or physically as in touch and hearing. Information is carried from the receptors, along the nerves to the brain.
- **Effectors** allow an animal's body to respond to stimulus from a nervous impulse. Vertebrate muscles are an example of effectors.

Glossary

Aqueous humour
A thin watery liquid between the lens and the cornea.

Balance
An animal's sense of equilibrium.

Cochlea
An organ of the inner ear that transfers vibrations from the ear bones to the nerves and on to the brain.

Cornea
The clear outer layer of the eye.

Eardrum
A tightly stretched membrane that vibrates as air molecules hit it.

Eye
A sense organ that detects light.

Hearing
A sense that records sound.

Iris
A mechanism that controls the amount of light entering the eye.

Lens
A jelly-like transparent substance that focuses the image on the retina.

Olfactory
Relating to the sense of smell.

Optic nerve
Connects the eye with the brain.

Photoreceptors
Special cells in the retina of the eye that respond to light.

Retina
A membrane containing cells that detect light.

Senses
Organs that allow an animal to find out about its environment.

Semicircular canals
Structures found in the ears that are essential to our sense of equilibrium.

Sight
A sense that records light.

Vitreous humour
Fluid inside the eye.

Science skills

Question 1

Which drawing best describes how you believe an eye detects light from a tree.

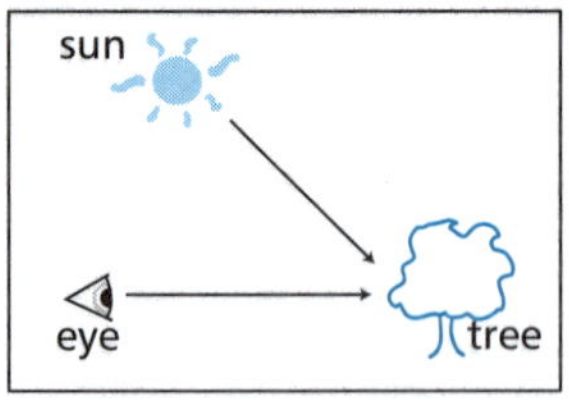

1 Light goes to tree and we look at tree.

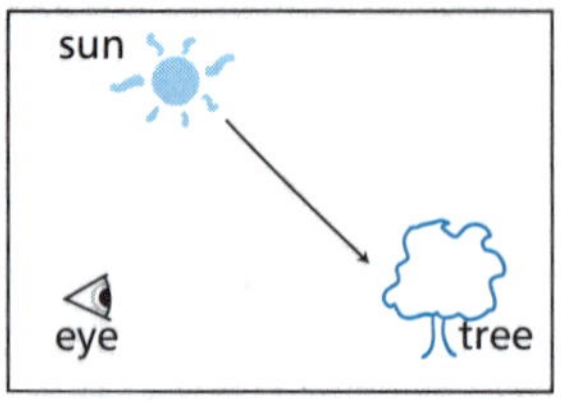

2 Light shines on tree and we see it.

3 Light is everywhere and we can see the tree.

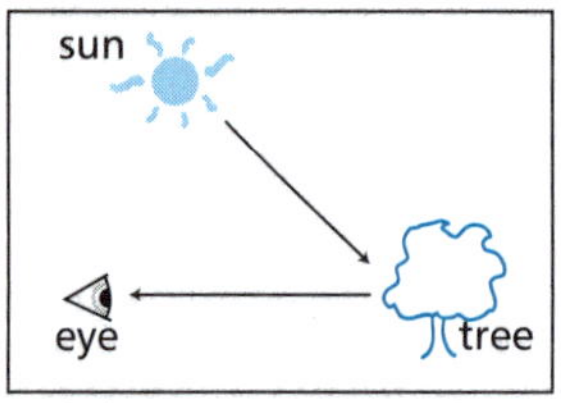

4 Light goes to tree and bounces to our eyes.

Question 2

Label the following diagrams with the correct description from the list below:

nearsighted
corrected nearsighted
normal near vision
farsighted
normal distant vision
corrected farsighted

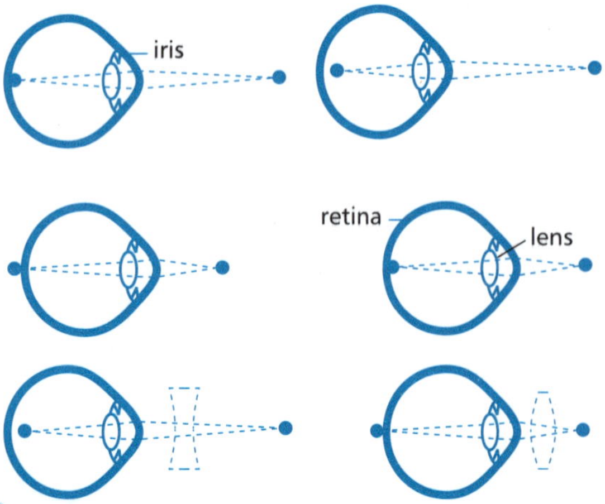

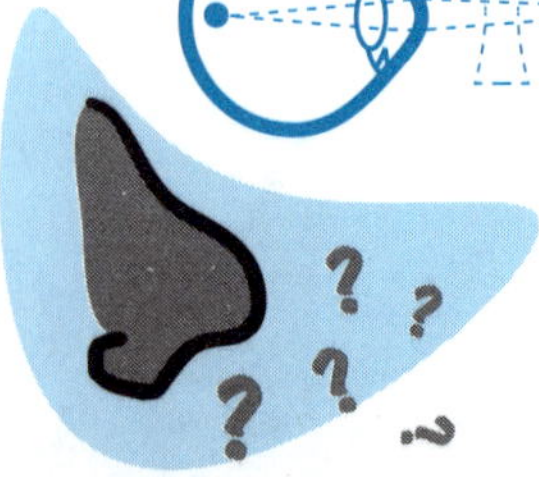

Question 3

Label the following structures of the eye:

pupil, lens, iris, cornea, retina, optic nerve

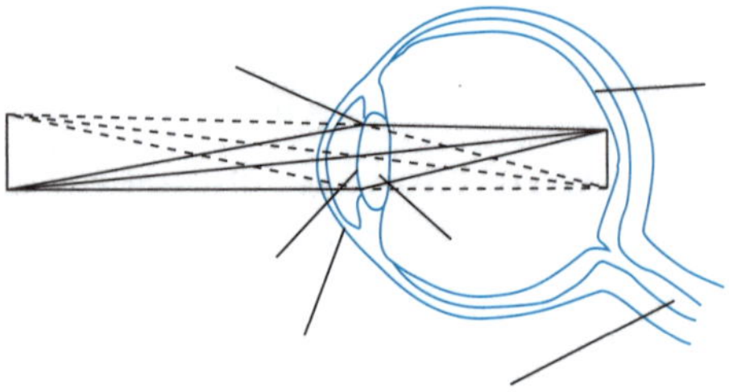

Question 4

Match the word or words in the column with the statement on the right.

a iris	location of visual receptors
b lens	carries nerve impulses from the retina to the brain
c optic nerve	focuses objects on the retina
d pupil	muscles control the pupil diameter
e retina	diameter decreases in bright light

Question 5

The eye works much like a camera.

Study the diagram below and match up the parts of the eye with similar parts of the camera.

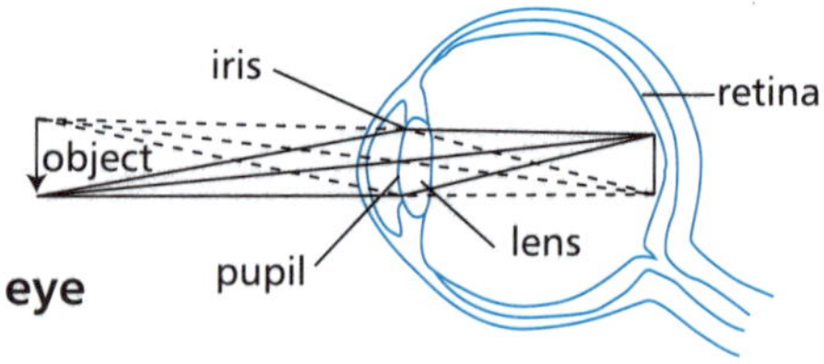

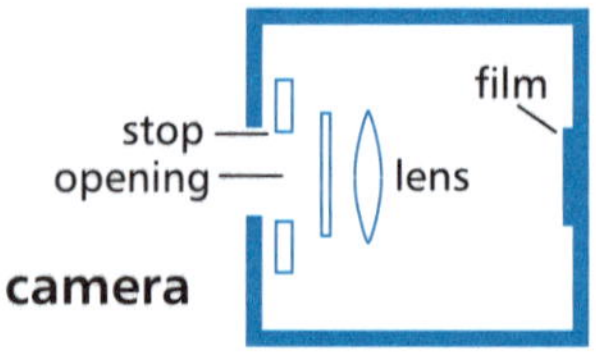

Question 6

A student uses a lens or magnifying glass to focus an image onto white paper. Will the image be upside down or right side up?

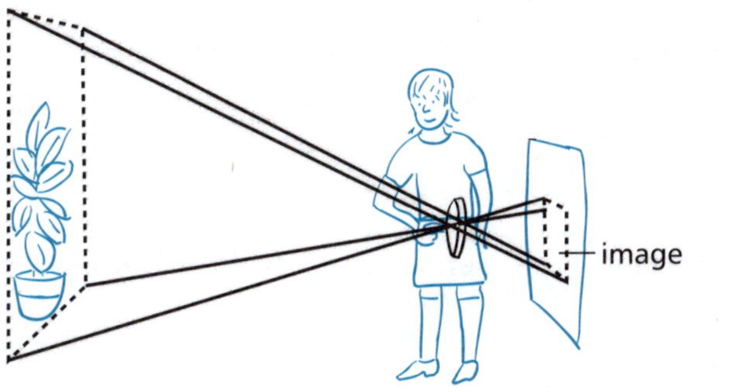

Question 7

Label the diagram of an ear below with the following words:

pinna, hammer, anvil, stirrup, semicircular canal, cochlea, eardrum, auditory canal

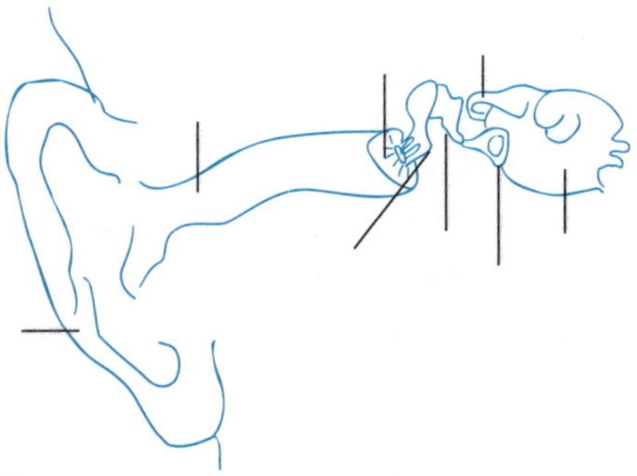

Question 8

The following statements explain how we hear.

Label the diagram below with the correct explanation.

A The movement of the ear bones causes motions in the fluid which fills the cochlea.

B The fluid moves the tiny hair cells which send nerve impulses to the brain.

C Sound waves enter the ear causing the eardrum to vibrate.

D The bones in the middle ear move back and forth.

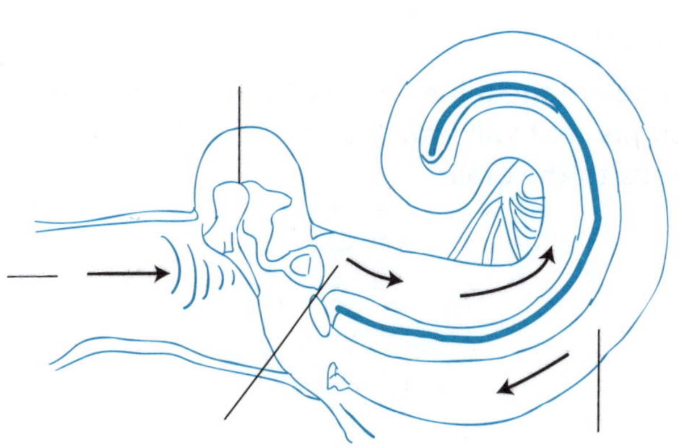

Question 9

Study the graph below and answer the questions.

a At what decibel level could sound be considered safe?

b What decibel output does a motorcycle have?

c At what decibel level do people with severe hearing loss begin to hear?

d Would people with mild hearing loss be able to hear a whisper?

Common levels of sound output	Decibels (dB)
heartbeat	10 dB
whisper	20 dB
optimum level for sleeping	35 dB
conversation	50–60 dB
annoyance level for normal hearing	75 dB
traffic sounds	80–85 dB
food blender	85 dB
discomfort level for normal hearing	95 dB
sewing machine	100 dB
radial saw	100 dB
motorcycle	110 dB
thunder	110 dB
jet take-off	120 dB
acute pain threshold for normal hearing	140 dB

Bars (left of table): Profound loss; Severe loss; Heard by people with moderate loss; Heard by people with mild loss; Heard by normal hearing

Question 10

Complete the table below with the area of the tongue that will signal the taste of the listed foods to the brain.

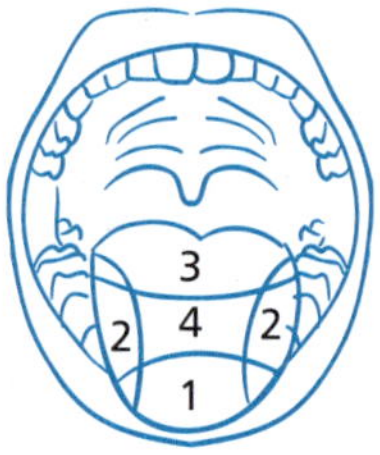

Substance	Area of tongue
sugar	
vinegar	
potato crisps	
vanilla extract (bitter)	
lemon juice	
honey	

Question 11

Students decided to test the distance between touch receptors in various parts of the body. They used two pins with different distances between them and recorded the closest distance at which two points could be detected.

a Predict the results of this experiment in the table below.

Area tested	Prediction
cheek	
forehead	
upper lip	
back of neck	
tip of finger	
back of hand	
lower arm	
upper arm	

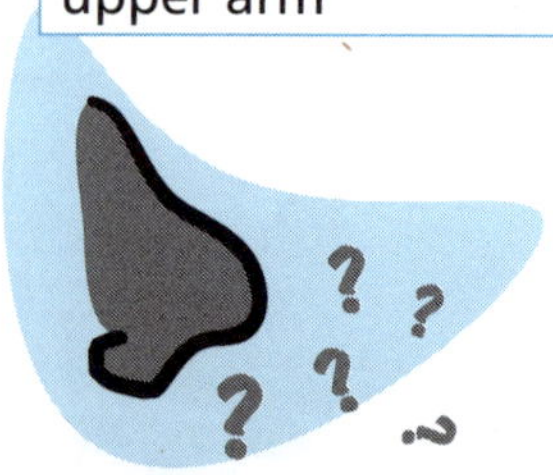

b What information did you base your predictions on?

Question 12

Animals use their senses for finding food. Animals' senses are adapted to the time of day that they hunt.

Complete the table below.

Animal	Senses used to find food
dingo	
bat	
owl	
sea eagle	

Question 13

a Find out which is your dominant eye by following the instructions below.

Instructions — Look at an object such as an upright piece of wood. With both eyes open, hold one of your fingers to line up with the object. Now cover each eye in turn. When you cover the dominant eye, your finger will suddenly seem to move, so that it is no longer lined up with the object.

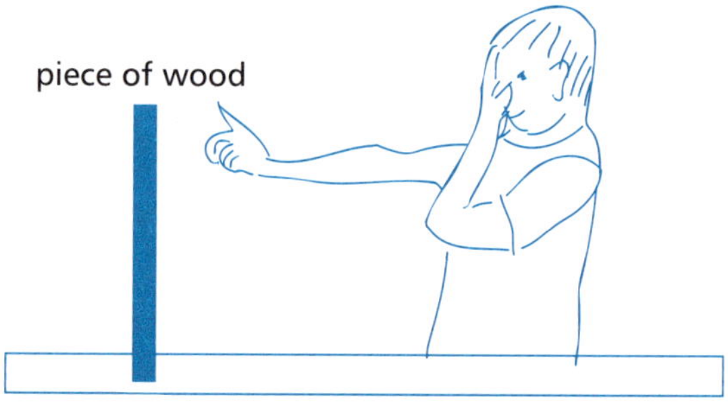

b Put the following explanations in the correct order and add the correct diagram reference in the space provided.

- Because your uncovered eye is not in line with your finger and the object, your finger seems to jump.
- When your dominant eye is covered, (diagram _____) your brain can only receive the image from the other eye.
- When your dominant eye is uncovered, (diagram _____) the piece of wood seems to line up with your finger.

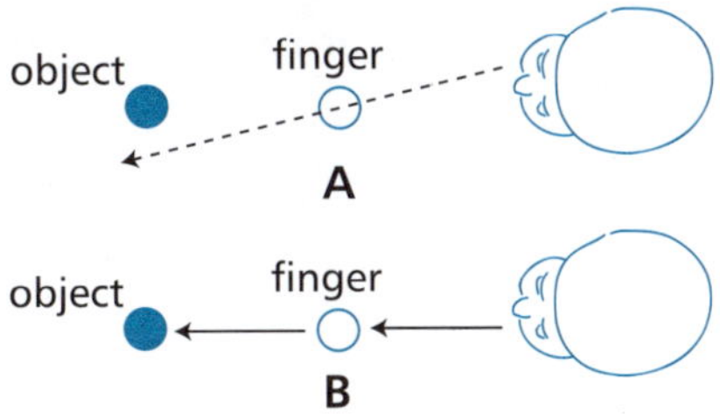

Question 14

Fill in the following flow chart with the sentences below to explain how a bat hears.

The bat hears these echoes through its sensitive ears.

The sound bounces back off any objects around it.

A bat produces bursts of high-pitched sound as it flies.

Bats use the echoes to pinpoint their prey.

↓
↓
↓

Revision questions

Question 1

a What is the function of the iris?

____

____

b If the retinal image is upside down, why do we see things right side up?

____

____

____

c What is the advantage of having two eyes?

____

____

____

Question 2

Complete the sentences below with the correct words.

Behind the __________ there is a convex __________ held in place by muscles. They can change the shape of the lens to allow you to __________ on objects close or far away. Between the lens and the cornea is a thin watery liquid called the __________ __________. Inside the eyeball is a fluid called the __________ __________.

Question 3

What is the name of the structure that maintains our balance?

____

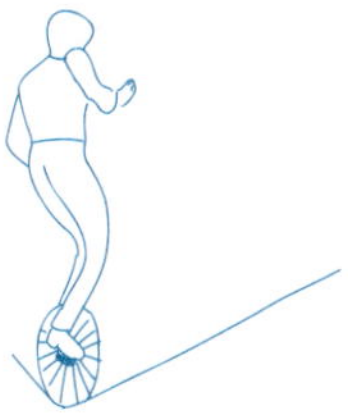

Balancing act
A tightrope walker must have a very good sense of balance.

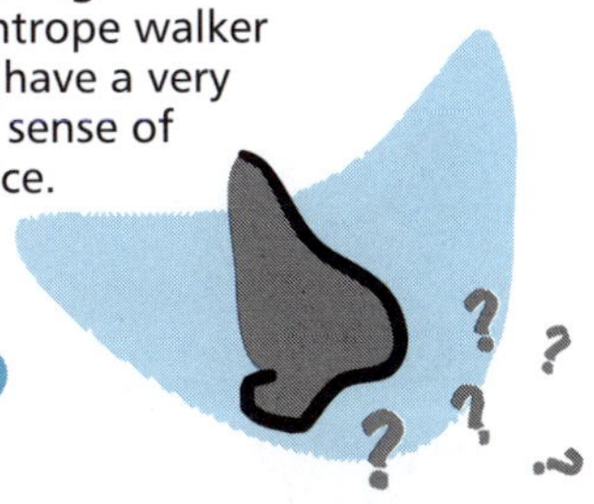

b People get dizzy when they spin around suddenly and then stop. Explain why this happens.

Question 4

List three ways in which the eyeball is protected from injury.

Question 5

Match the following:

Taste	a sheet of cells used to detect airborne chemicals
Taste bud	a sense that detects dissolved chemicals
Smell	a small organ that detects dissolved chemicals
Olfactory membrane	a sense that detects airborne chemicals

Question 6

a In bright light the muscles in the iris ____________ and narrow the pupil to ____________ the amount of light striking the retina. In dim conditions the muscles in the iris ____________ and the amount of light entering the eye ____________.

b Label the following diagrams with either 'dim light' or 'bright light'.

A

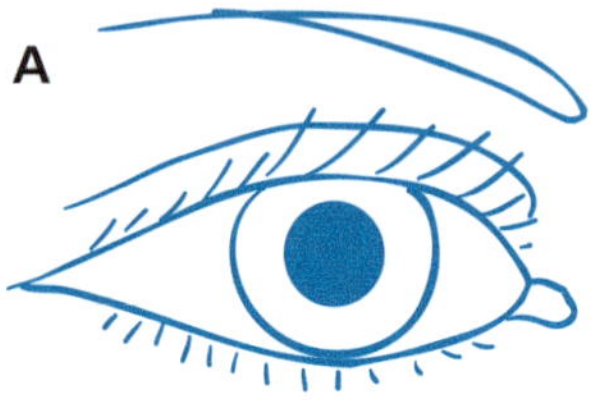

B

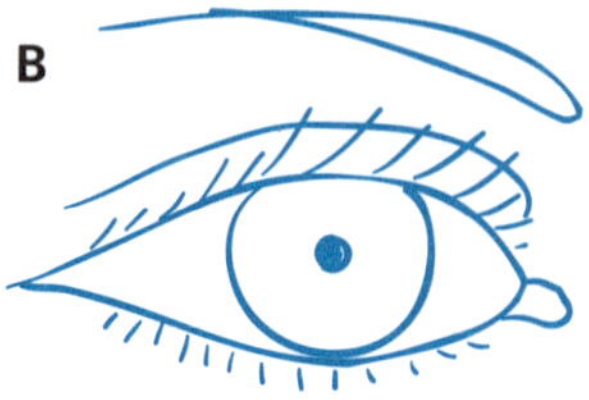

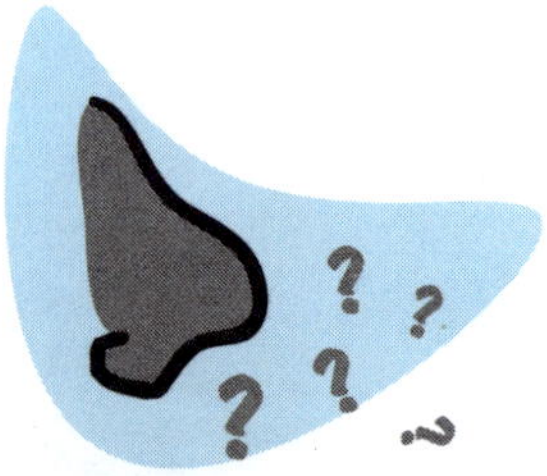

Question 7

In the table list the main parts of the ear.

Outer ear	Middle ear	Inner ear

Question 8

a What is a stimulus?

b What is a receptor?

c What is a response?

d Are the receptors for 'sweetness' located in the same area of the tongue for everybody?

e Where are 'sour' receptors located?

f What are the four basic tastes that tongue receptors can distinguish?

Question 9

Human feeling for temperature is unreliable — it depends on whether our bodies are feeling hot or cold.

A student set up the following equipment. Then he soaked one hand in the hot water and one in the cold water. Next he put both hands in the warm water.

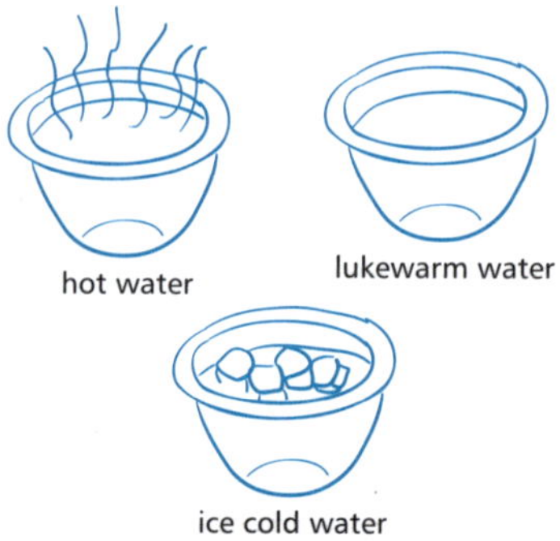

a Which hand felt hot and which felt cold?

b Explain why this sensation happened?

Question 10

We cannot always trust our sense organs. The following diagrams are optical illusions, which means that they trick our sense of sight.

Answer the questions under each diagram.

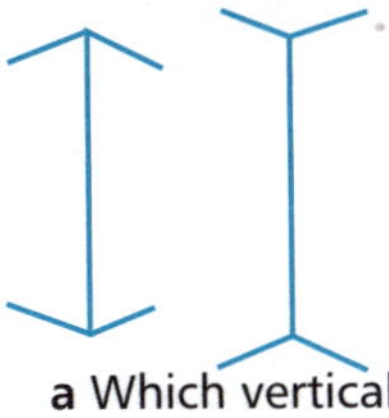

a Which vertical line is longer?

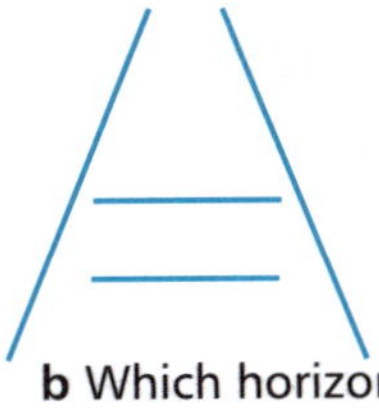

b Which horizontal line is longer?

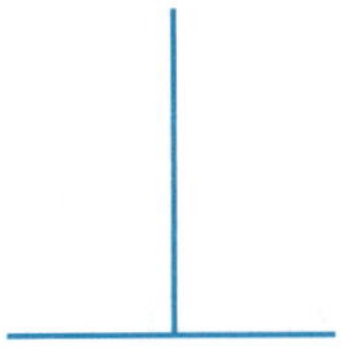

c Is the vertical or horizontal line longer?

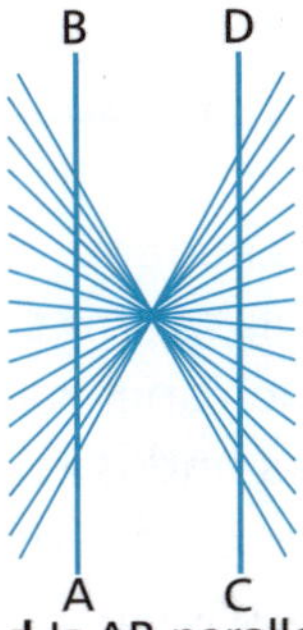

d Is AB parallel to CD?

Question 11

Sensory impulses from the nose travel to the brain by way of the:

a cranial nerve

b spinal nerve

c olfactory nerve

d optic nerve

Question 12

The light sensitive part of the eye is called the:

a vitreous humour

b cornea

c retina

d iris

Question 13

Which of the following animals have an outer external ear (pinna)?

a snakes

b frogs

c shark

d bandicoot

Question 14

The senses respond to either chemical or physical stimulus. Complete the table below.

Sense	Stimulus
vision	
hearing	
touch	
smell	
taste	

Question 15

Two students investigating the function of the ears carried out an experiment where one student rings a bell in different positions near a blindfolded student.

a How does a pair of ears accurately detect the direction from which sound waves are coming?

b From which position in the room do you get the best result when you listen to your stereo?

CHAPTER 3

ANIMAL REPRODUCTION

Multiplication is the name of the game — but why?

All living things have a limited lifespan, so they must reproduce to ensure the continued existence of the species. All animals can increase their numbers but things like the availability of food and predators keep their numbers in check. **Reproduction** is the production of offspring.

Animals have developed an incredible variety of ways to reproduce. However they all follow one of two basic plans:

- Animals that can reproduce without having to mate (asexual reproduction).
- Animals where two parents must mate to produce a fertilised egg (sexual reproduction).

Asexual reproduction

This form of reproduction only needs one parent. Many simple animals reproduce without mating. Some simply bud off small parts of themselves while others develop from unfertilised eggs. Asexual reproduction is usually simple and fast and produces offspring with exactly the same genetic blueprint or **genotype** as the parent. The offspring is therefore a **clone** — a cell, a group of cells, or an organism with exactly the same genes as the parent. The offspring grows up to have the same characteristics the parent.

Sexual reproduction

In sexual reproduction there are always two parents. The sex organs, called gonads, of each parent produce **sex cells** or gametes. When a male and a female sex cell come together, in a process called **fertilisation**, they combine to form a fertilised egg called a **zygote**, which develops into a new individual.

Offspring produced by sexual reproduction each have a unique collection of genes. They are not clones of one of their parents. This genetic diversity gives these species a better chance of survival.

The male sex cell

Male sex cells are called **spermatozoa** or sperm. They are produced in an organ called the testis, by a process called **spermatogenesis**. Each sperm contains slightly different genetic information from the other sperm. A sperm has a head and a tail that whips from side to side to push the cell along. Each sperm contains one set of chromosomes and is termed **haploid**. Sometimes millions of sperm are produced to go searching for ovum but usually only one is successful in bringing about fertilisation.

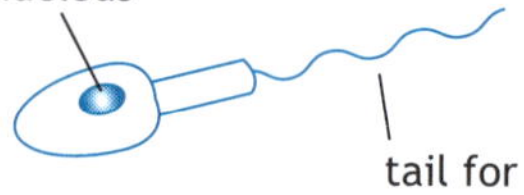

A male sex cell

The female sex cell

The female sex cell is called an **ovum** (plural ova) and it is also haploid. Each egg is a genetic individual. They are produced in an organ called the ovary by a process called **oogenesis**. They are usually larger than sperm and consist of a nucleus and a substance called yolk, which is a food source. In mammals the yolk is small but in birds and reptiles it is larger. The ova of mammals are formed before birth and are released at a steady rate throughout adult life.

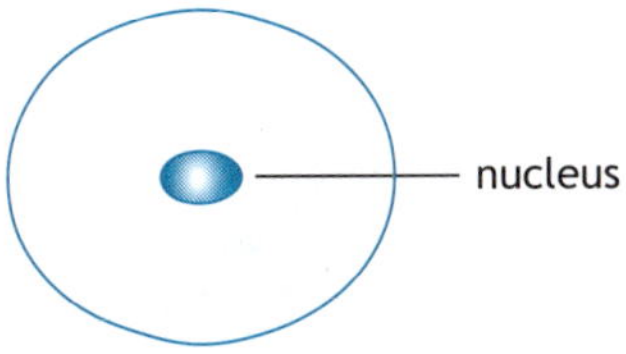

A female sex cell

Fertilisation

During fertilisation a male sex cell binds to the outside of an ovum and the membranes join together. The sperm's nucleus then enters the ovum and fuses with the ovum's nucleus. This results in a fertilised cell or zygote. The zygote contains two sets of chromosomes, one from each parent and is termed **diploid**.

Sexual reproduction evolved in seawater as this is a friendly environment for gametes and fertilisation can be achieved simply by the male releasing sperm into the water that already contains the eggs. When fertilisation occurs outside the body it is called **external fertilisation**. Animals that have learned to live on land carry out **internal fertilization**, which occurs inside an animal's body. This is a way to ensure that the gametes do not dry out.

Vertebrate reproduction

The fertilisation and development of fish is still linked strongly to water. Development is rapid and largely external. The yolk of the fish egg sustains life for a short period and then the young fish must fend for itself. However numbers are high and the death rate is also high.

Land vertebrates needed considerable innovation in the process of reproduction and development to prevent their gametes drying out.

- **Amphibians** — reproduction is still linked closely to water. Fertilisation is external and large numbers of eggs are often laid in water. Development is slower and often involves two stages, for example: egg → tadpole → frog.
- **Reptiles** — the aquatic habitat has been abandoned and eggs are fertilised inside the mother. An organ called a penis is used to inject the male sex cells inside the female. Fewer eggs are produced, and they are laid in the soil and covered in a tough leathery case to prevent drying out, before hatching independent of the parents.
- **Birds** — internal fertilisation occurs and the eggs have a hard outer case to prevent them drying out. The parents warm the eggs and nurture the young to increase the chances of survival.
- **Mammals** — fertilisation is internal and a massive network of blood vessels called the **placenta** provides nourishment inside the body so the young foetus does not dry out.

  Marsupials nourish their young in a pouch until they are big enough to survive in the outside environment.

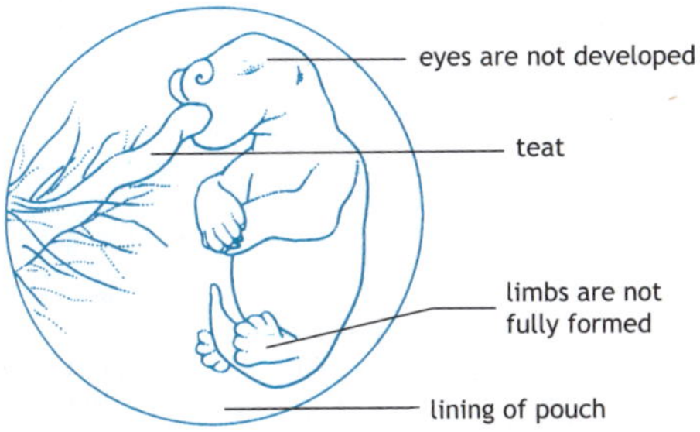

A young wallaby

- **Monotremes** — (platypus and echidnas) lay leathery eggs, incubate them and produce milk via a mammary gland in the skin.

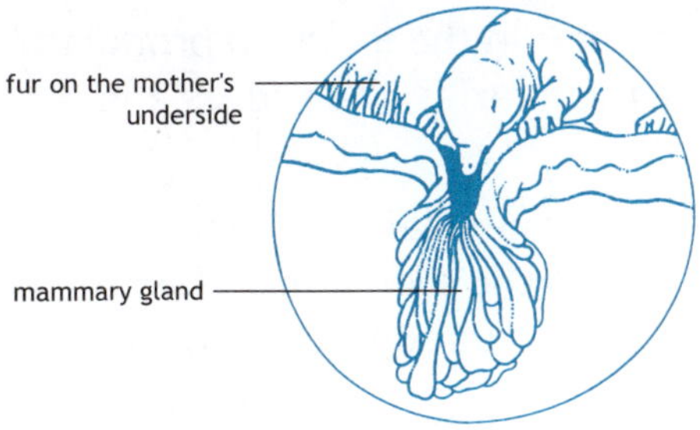

A young echidna

The human reproduction system

Unlike many animals, placental mammals such as humans reproduce very slowly and the embryo stays inside the mother's body until it is well developed.

The male reproductive system

The main organs making up the male system are two **testes** and the **penis**. The sperm are made by the testes, inside tubes called seminiferous tubules and then they move to the epididymis where they mature. The testes also produce the male sex hormone **testosterone**.

The penis is the organ that transfers the sperm into the female's reproductive tract or vagina.

During intercourse, muscles pump sperm from the epididymides along a tube or sperm duct called the vas deferens. The sperm is mixed with a nourishing fluid and this seminal fluid, the semen, containing the sperm is ejaculated from the penis into the female.

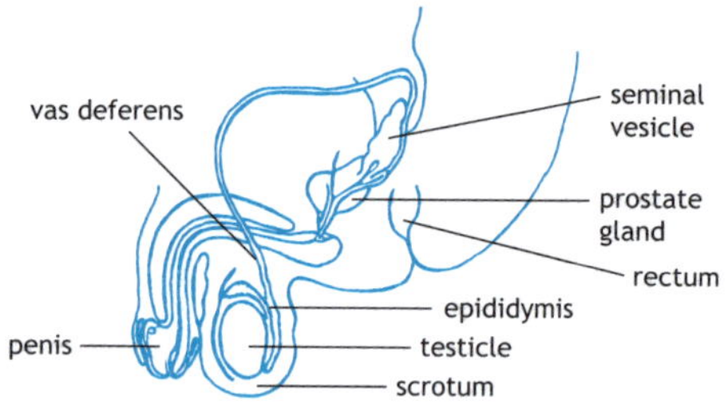

The male reproductive system

The female reproductive system

The main organs in the female reproductive system are the **ovaries**, the **uterus** and the **vagina**.

The vagina is a muscular tube that leads to the uterus. It carries the sperm and when the baby is born it comes out through the vagina.

The uterus is the muscular chamber in which a baby develops. At the top are two fallopian tubes that lead to the ovaries. The cervix is the opening at its base that leads to the vagina.

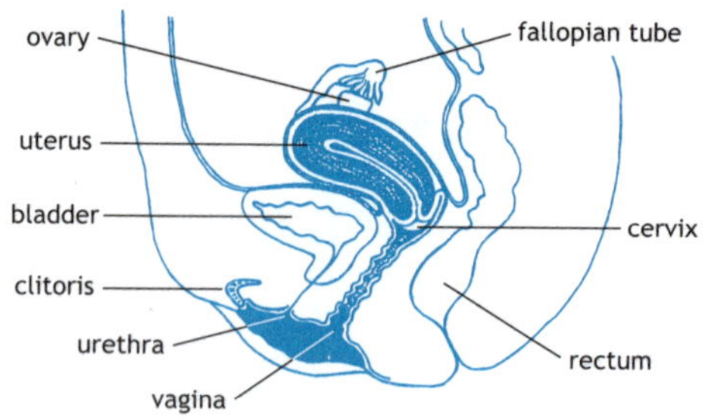

The female reproductive system

Pregnancy and birth

During ovulation an ovum is released from an ovary and the uterus thickens in preparation for pregnancy.

- Ovulation occurs during a process called the menstrual cycle, which lasts about 28 days. The ovum passes down the fallopian tube towards the uterus. If the ovum is fertilised it implants itself in the uterus lining. If the ovum is not fertilised the lining of the uterus is shed during a process called menstruation (monthly period) and the cycle begins again.

  This process is controlled by chemicals called hormones that are made by the ovaries. **Oestrogen** regulates ovulation and **progesterone** prepares the uterus for a fertilised ovum. If fertilisation does not occur it breaks down.

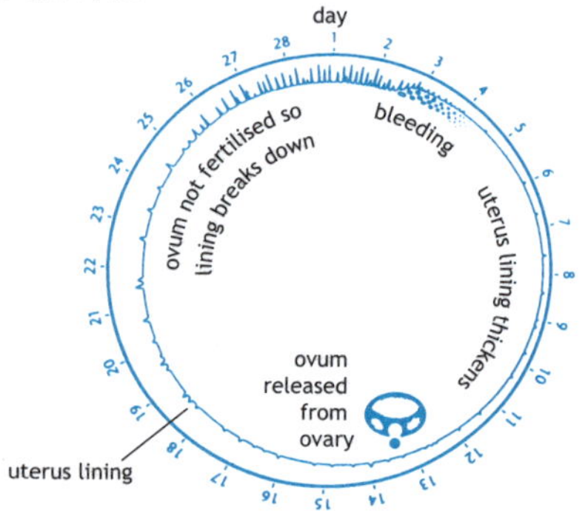

The menstrual cycle

- An **ovum** is a single cell, which after fertilisation, starts to divide forming a tiny ball of cells called a blastocyst that becomes surrounded by a layer of uterine cells. This ball of cells grows into an **embryo** and after about eight weeks it is called a **foetus**.

- The **placenta** is a spongy mass of tissue that attaches to the foetus. It is in the placenta that the bloodstreams of the mother and foetus come close together, without actually mixing. It allows the passage of oxygen, carbon dioxide, and dissolved food substances to pass between mother and baby. The baby's food reaches the placenta via the **umbilical cord**.
- Before **birth** the baby turns around so that its head points downwards towards the cervix. A hormone called oxytocin causes the uterus to contract. These contractions become stronger and eventually push the baby's head through the cervix and vagina and out of the mother's body.

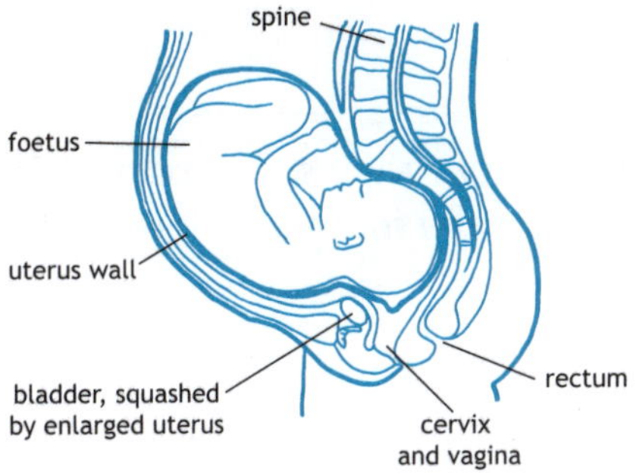

Pregnancy

- Once the baby has emerged, it takes its first breath and blood begins to circulate. The umbilical cord is cut and tied off and a little later the placenta or afterbirth is expelled.

Glossary

Asexual Reproduction
Reproduction involving only one parent. The young have exactly the same chromosomes as the parent, and therefore the same characteristics.

Birth
The expulsion of a baby from the uterus.

Clone
Genetically identical individuals or cells.

Diploid
Contains two sets of chromosomes, one set from each parent.

External fertilisation
Fertilisation occurring outside the parents' bodies.

Fertilisation
In sexual reproduction, the stage at which male and female sex cells join together to form a single cell.

Gamete
A haploid sex cell. Male and female gametes unite during sexual reproduction to produce a diploid zygote.

Genotype
The genetic make-up of an individual.

Gonads
The male and female sex organs.

Haploid
A cell containing only one set of chromosomes.

Hormone
Hormones are chemical messengers that travel through the body in the blood until they reach their target cells.

Internal fertilisation
Fertilisation occurring inside a parent's body.

Oogenesis
The process in the ovary that results in the production of female gametes.

Ovulation
The release of the female sex cell from the ovaries.

Ovum
The female gamete, the haploid; an unfertilised egg.

Ovary
The part of an animal or plant that produces the female sex cells.

Penis
The male organ that transfers the sperm into the female's body.

Placenta
A complex layer through which food passes from the mother's blood to her unborn young.

Reproduction
The production of offspring.

Sex cells
The female ovum and the male sperm.

Sexual reproduction
A way of producing young that needs two parents. In sexual reproduction the young get half of their chromosomes from each parent and so receive some of their characteristics from each of them.

Spermatozoa
Male sex cells.

Testis
An organ in males that produces sperm.

Uterus
The organ in females in which a baby develops.

Vagina
The start of the female reproductive tract where the sperm are deposited.

Zygote
The diploid of the union of haploid gametes in fertilization; a fertilised egg.

Science skills

Question 1

Placental mammals have a special way of nurturing their young while they are still inside the mother's body.

a Explain this process.

____

____

____

b Which structure allows transport of food and oxygen from the mother's blood to the embryo?

____

c Label the diagram below of a developing mammalian embryo. Use the following terms:

uterus, placenta, lining of the uterus, embryo, umbilical cord.

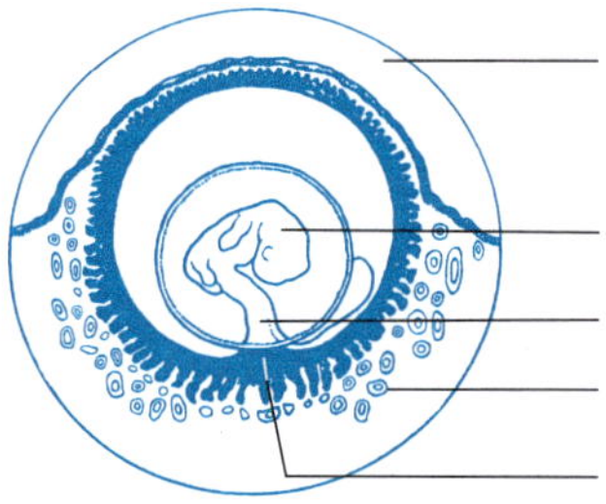

Question 2

On the following diagrams label the correct structures with the words provided below.

fallopian tube, testis, vagina, oviduct, vas deferens, cervix, uterus, penis, ovary.

Female reproductive organs

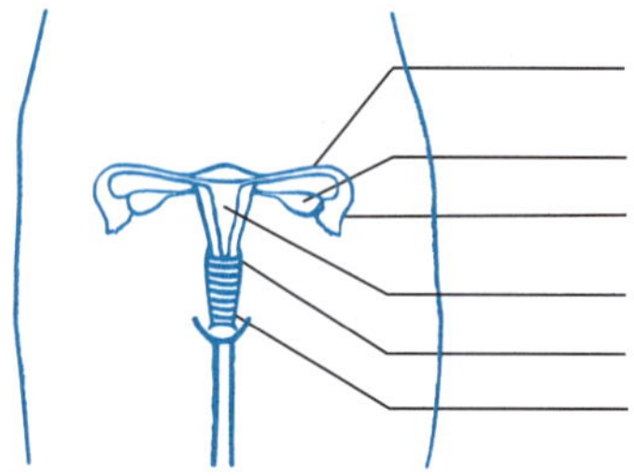

Male reproductive organs

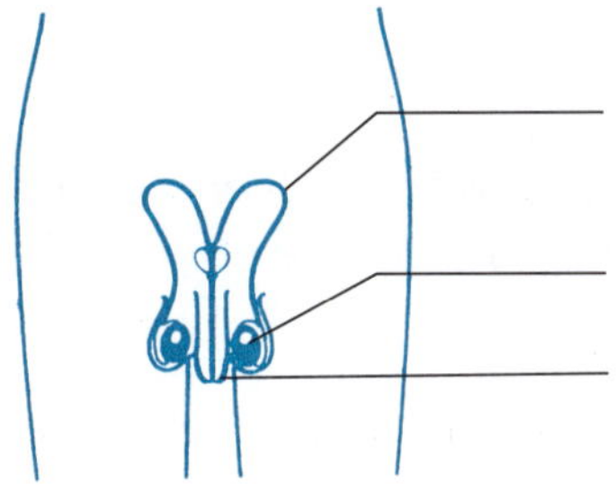

Question 3

The female menstrual cycle lasts for 28 days during which an ovum is released from one of her ovaries.

The following time line represents the menstrual cycle:

Week 1 Week2 Week 3 Week 4

_______|_______|_______|_______

day 1 day 14 day 28

Organise the following statements into the correct cycle of events and match to the correct week.

- If the ovum is not fertilised it dies.
- During the period of bleeding the lining of the uterus is shed.
- The lining of the uterus continues to thicken.
- A new lining grows inside the uterus ready for a fertilised ovum to grow.

Question 4

The following table provides a list of organs. Tick the box to indicate whether it is a male or female part.

Organ	Male	Female
fallopian tube		
seminal vesicle		
prostate gland		
cervix		
vas deferens		
ovary		
epididymis		

Question 5

The table below compares sexual and asexual reproduction. Mark the correct box with a tick.

Characteristic	Asexual	Sexual
involves mating		
involves one parent		
produces a zygote		
different genotype from both parents		
involves two parents		
involves sex cells		
has the same genotype as the parent		
produces a clone		

Question 6

Write the correct vertebrate group into the space beside the description.

Description	Vertebrate group
nourish their young in a pouch	
internal fertilization; parents abandon leathery coated eggs in the soil	
eggs have a hard coat and parents nurture the young	
internal fertilisation and nourishment; mother produces milk	
large numbers of eggs laid in water; development is external to the parents body	
reproduction linked to water, fertilisation is external, adults live on land	

Question 7

a To which group of vertebrates does the following animal belong?

____

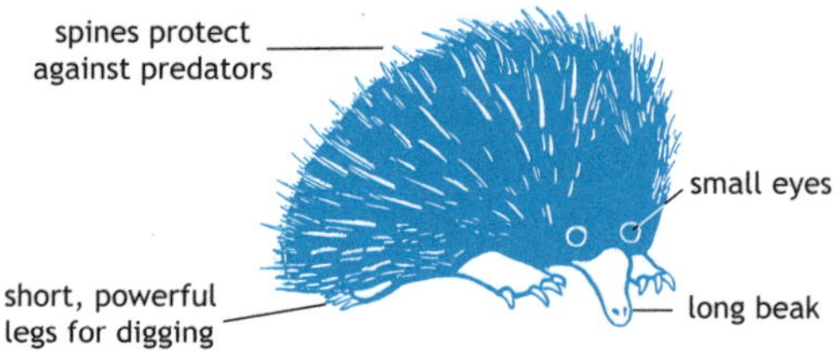

b How do they reproduce?

c Do they suckle their young?

d Name another Australian animal that belongs to this group.

Question 8

Label the diagram below with the following terms:

umbilical cord, placenta, fetus, cervix, uterine wall

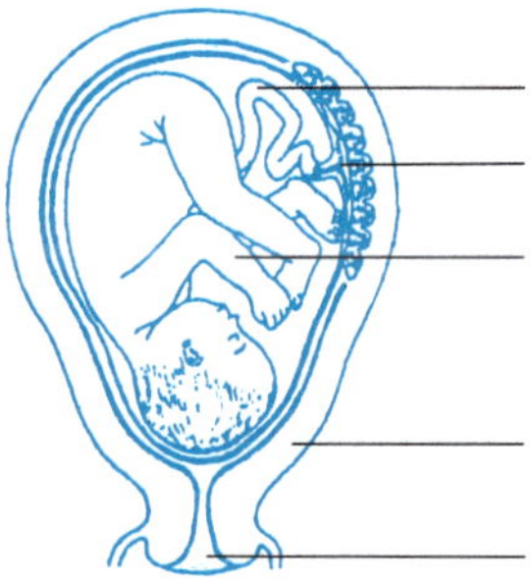

Question 9

Study the graph below and then answer the questions.

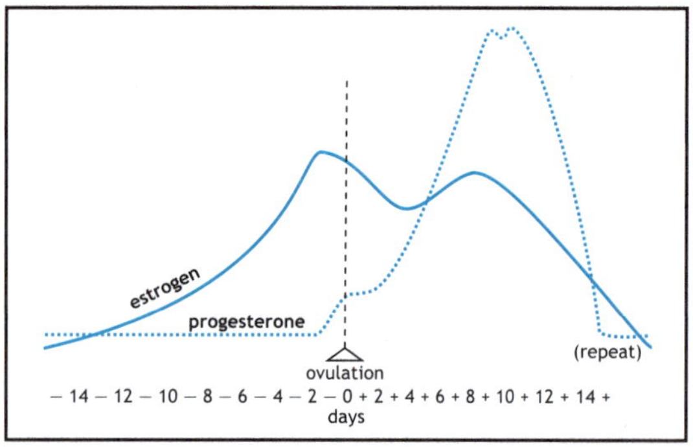

a Why does progesterone increase after ovulation (day 0)?

b Why do progesterone levels fall at about day 12?

c What happens to the lining of the uterus once blood progesterone levels have fallen?

d At what point do oestrogen levels reach their highest point?

e What is a woman's period?

f What would happen to progesterone levels if the female were pregnant?

Question 10

When studying reproduction in placental mammals, four students were given a list of organs and what their respective functions were. They matched them as in the table below. Which student matched them correctly?

	Oviduct	Penis	Uterus	Testis
1	ovum production	sperm transfer	embryo development	sperm production
2	fertilisation	sperm transfer	embryo development	sperm production
3	embryo development	sperm production	fertilisation transfer	sperm
4	fertilisation	sperm production	embryo development	semen production

Question 11

Study the following table:

Characteristics	Species X	Species Y	Species Z
adult habitat	terrestrial	aquatic	terrestrial
embryo	in an egg	in water	swim in
development method of breathing	lungs	gills	water lungs
mating environment	land	water	water

In which species is internal fertilisation likely to occur?

a X only **b** Y and Z **c** X and Y **d** X and Z

Question 12

Study the diagram below and mark with the correct letter the site of the following processes.

a fertilisation **b** sperm entry

c foetus development

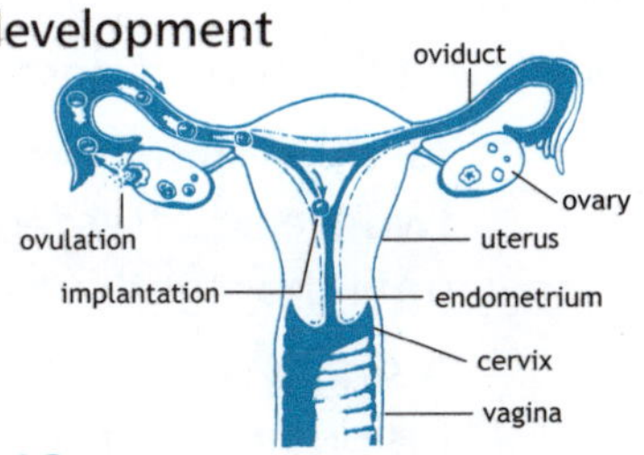

Question 13

Add the correct organ to match the function.

Function	Organ
the fertilised egg develops here	
sperm is put here	
male sex cells travel along here to meet the egg	
makes the ovum	

Question 14

Study the following diagram.

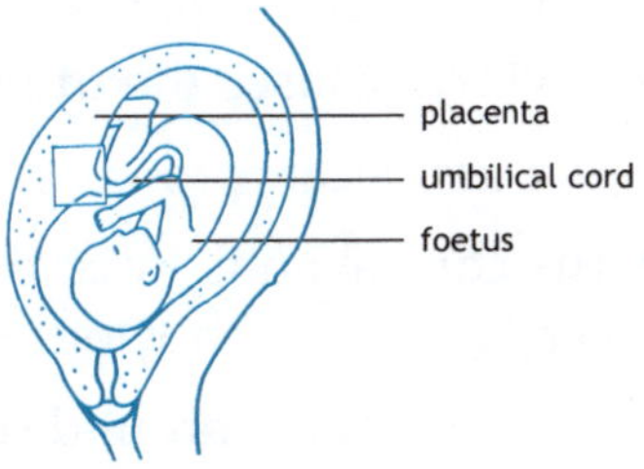

a What special organ is formed when an embryo embeds itself in the uterus lining?

____

b What connects the placenta to the foetus?

____

c What two substances pass from the mother to the foetus through the placenta?

____

d What two substances pass from the foetus to the mother?

____

e Is the placenta needed after the birth of the baby?

____

Revision questions

Question 1

Why do amphibians lay their eggs in water?

____

____

Question 2

Placental mammals have fewer young than other vertebrates because more of them survive. Why is this so?

____

____

Question 3

Fill in the missing words to complete the sentences below.

To make a new individual we need two special cells called ______ cells. The female cell is called an ______ and the male is called a _______. These calls have only one set of genetic material and are called ________ cells.

Question 4

For a new human to start to grow, a sperm and an ovum must join up.

a Label the following diagram with the words: sperm, ovum, zygote, haploid and diploid.

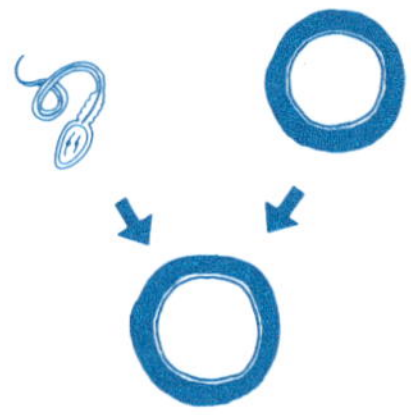

b What is fertilisation?

c How do sperm get to the fallopian tube from the opening of the uterus to fertilise the ovum?

d Which organ is used by the male to deposit the sperm in the female vagina?

Question 5

a What are the two main jobs for testes?

b What is the male hormone called?

c Which sex cell is the biggest, the male or the female?

d Where do sperm mature?

e What is the name of the sperm duct?

Question 6

a What is a hormone?

b List two female reproductive hormones.

c Name one male reproductive hormone.

d Which hormone is responsible for the preparation of the uterus for implantation of a zygote?

Question 7

Match the word with the correct description:

Zygote	All development stages from two weeks after fertilisation until the eighth week.
Blastocyst	A single cell resulting from the fusion of a sperm and ovum nucleus.
Foetus	A solid ball of cells.
Embryo	All development stages from the ninth week until birth.

Question 8

The birth process follows several stages. Rearrange the following steps to accurately describe the birth of a baby.

- Oxytocin causes the uterus to contract.
- The umbilical cord is cut and tied.
- The contractions force the head through the cervix and the baby out of the mother's body.
- The baby turns around so that its head points towards the cervix.
- The afterbirth is expelled.
- The baby takes its first breath and blood begins to circulate.

Question 9

In the formation of a new organism by sexual reproduction, which of the following is the correct sequence of stages?

a sex cells, fertilisation, zygote, embryo

b fertilisation, zygote, sex cells, embryo

c sex cells zygote, fertilisation, embryo

d fertilisation, embryo, sex cells, zygote

Question 10

Why do birds lay eggs?

____

____

Question 11

Match the hormone with its function.

Oestrogen	Promotes secondary sex characteristics in males.
Oxytocin	Maintains a thickened uterine lining.
Testosterone	Promotes secondary sex characteristics in females and monitors ovulation.
Progesterone	Produces contractions of the uterus during birth.

Question 12

Fill in the missing words:

The menstrual cycle is a way of making sure that the lining of the _______ is ready for the implantation of an embryo each time an _________ releases an ovum. The menstrual cycle is controlled by _____________.

Question 13

What are the two main functions of the female reproductive system?

____

____

____

____

Question 14

Study the diagram below and write down the name of the type of fertilisation occurring.

____

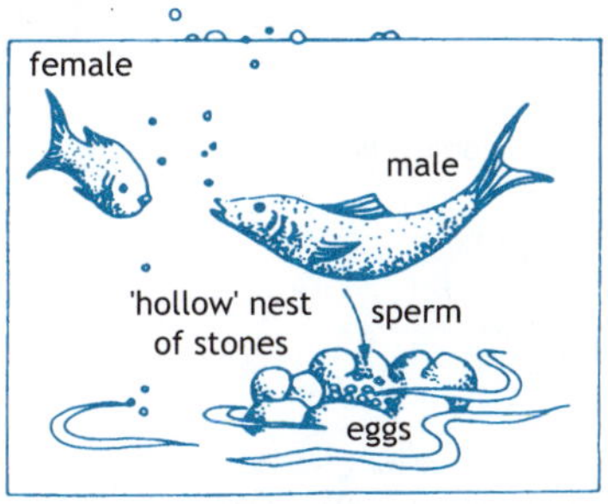

The male stickleback squirts his sperm over the eggs laid in the nest by the female.

Question 15

a What is the difference between gametes and gonads?

____

b How do hormones travel around the body?

____

c What tube connects the testis to the penis?

____

d What other substance is added along this tube?

____

e Why must the uterus have a muscular wall?

____

____

Question 16

The diagram below shows two earthworms mating.

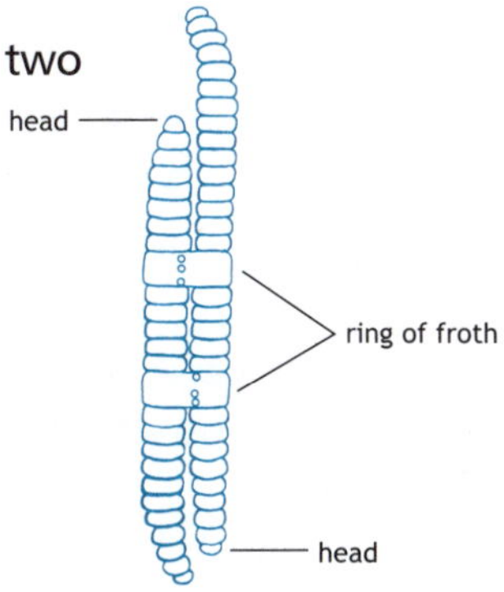

Sperm is passed from one to the other. Is this external or internal fertilisation?

____

Question 17

The following diagram is of an anemone that is budding off a new individual from the side.

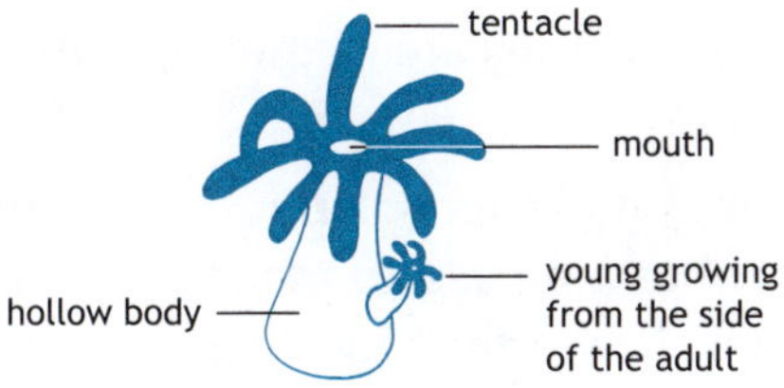

What kind of reproduction is this?

____

Question 18

Fertilisation starts when the _______ and the ovum join. The __________ divides as it travels down the oviduct to the __________. This __________ sinks into the lining of the uterus where it develops into an __________.

CHAPTER 4

INFECTIOUS DISEASES

Who will win the war?

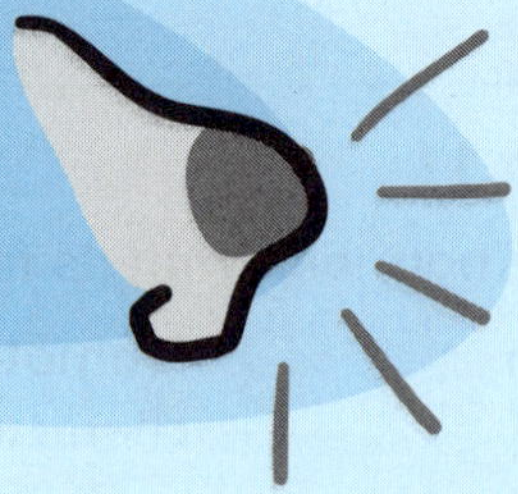

An animal's body is a biological battleground. Micro-organisms settle on it and try to break in, while the body retaliates and tries to fend them off.

An animal's body is usually kept in a stable condition by the control system called **homeostasis**. This stable state sometimes breaks down, and the body stops functioning normally. This breakdown is called a **disease**.

An infectious disease, such as measles, is one that is triggered by an infection, or by a growth of **pathogens**. A non-infectious disease is triggered by other factors such as inherited characteristics. Pathogens enter the body through the skin, through the mouth or by respiration (breathing them in).

What is a microbe?

A microbe is usually just a single cell and so cannot be seen without a microscope. Microbes are all around us. They are in the air, water, food and our bodies.

Some microbes make us sick but others are very useful. Microbes are used to make bread and cheese; others live in the digestive systems of some animals and help digest their food. Microbes break down dead animals and plants and some are used to process sewage. These are called saprophytic bacteria. Some diseases like malaria are caused by **parasites**. A parasitic organism depends on the host and the **host** is harmed.

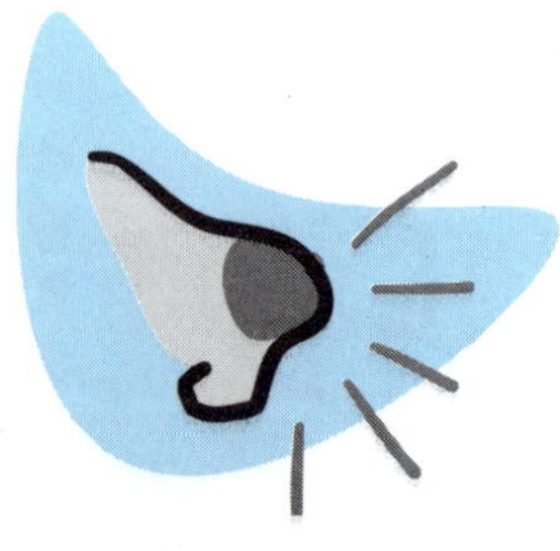

Microbes that cause disease

The most common microbes or pathogens causing disease are bacteria and viruses. However some fungi and unicellular organisms are also important.

How big are bacteria and viruses?

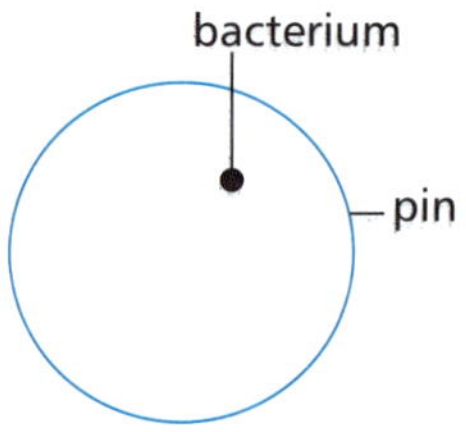

One of the biggest types of bacterium compared with the head of a pin.

Size of bacteria

Pathogenic bacteria

Pathogenic bacteria damage the cells of animals and plants causing disease. Sometimes they produce poisonous wastes or toxins. If bacteria get into your sweat glands a pimple may form.

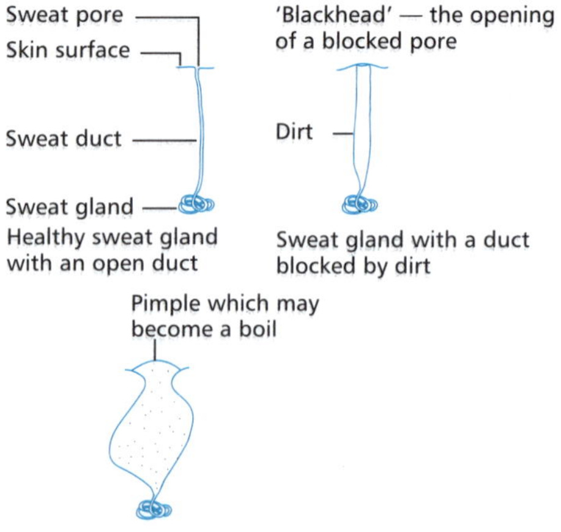

Infected sweat gland. The duct is filled with dirt and 'pus', which is the accumulated bacteria.

Bacteria absorb nutrients and divide and multiply like all normal cells. Diseases caused by bacteria can spread to other organisms very quickly. Examples of diseases caused by bacteria are listed in the table below:

Disease	What it does	How it spreads
tonsillitis	headache, sore throat, raised temperature	sneezing, coughing, spitting
whooping cough	bad cough, raised temperature	sneezing, touching, spitting
salmonella food poisoning	diarrhoea, feeling sick	eating infected food
tuberculosis	very ill, a bad cough	water droplets in the air

'Cocci' are round-shaped bacteria.

'Bacilli' are rod-shaped bacteria.

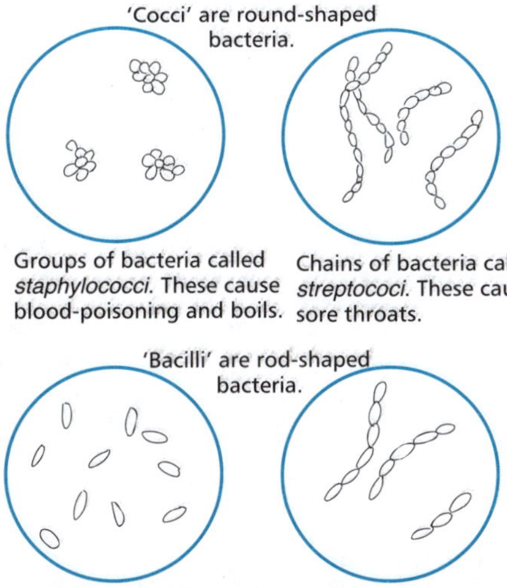

Groups of bacteria called *staphylococci*. These cause blood-poisoning and boils.

Chains of bacteria called *streptococi*. These cause sore throats.

Single rod-shaped *bacilli*. These cause tuberculosis.

Chains of rod-shaped *bacilli*. These cause dysentery.

What bacteria look like

Viruses

Viruses are bundles of genetic material that attack cells and take chemicals from them. They cannot exist on their own. Each type of virus causes a different disease. Some are easier to catch than others and sometimes you don't catch them again after you have had them once.

Examples of viruses follow:

Disease	What it does	How it spreads
influenza	aches, pains, runny nose, high temperature	water droplets in the air
measles	small red spots and a skin rash	close contact
chicken pox	raises itchy spots on skin	close contact
mumps	swollen cheeks and neck	close contact
AIDS	destroys some of the cells of the immune system	sexual contact or blood
Ross River fever	fatigue and high temperature	blood-sucking insects

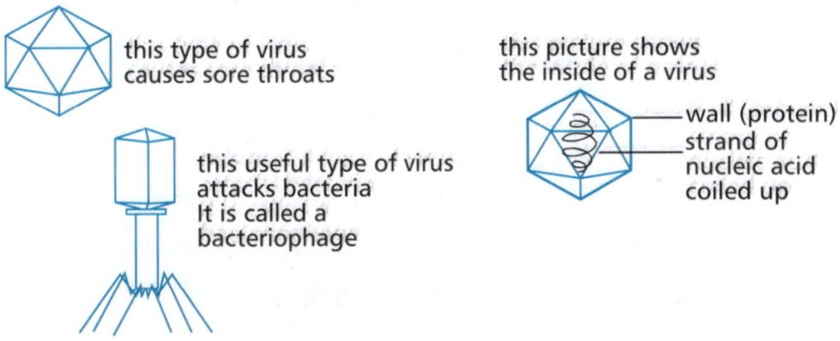

What viruses look like

Fungi

Fungal infections usually attack the skin and do not invade deeper layers. Athlete's foot is a disease that causes itching between the toes. It is spread by bare feet coming into contact with infected areas. Ringworm is a fungus present in dogs and cats and is transmitted by contact with damaged or infected skin

Parasites

A parasite lives in or on another organism. **Vectors** carry parasites and pathogens from one host to another, for example, the malaria mosquito carries plasmodium from one person to another.

Malaria mosquito

Disease	What it does	How it spreads
malaria	severe recurring fever	mosquito
typhus	fever, muscle pain, rash	lice, mites, ticks, fleas
amoebic dysentery	diarrhoea, blood in the faeces	protozoan in food and water
bilharzia	rash, cough, diarrhoea	water snail

Killing pathogens

- **Bacteria** and fungi in food can be prevented by using heat (sterilising), cold (freezer), sugar (jams), salt (bacon), weak acid like vinegar (pickled onions), or drying (dried fruit).
- **Antiseptics** and **disinfectants** are chemicals applied to the skin surface that kill bacteria.
- **Antibiotics** should kill bacteria if taken according to the directions, but not viruses. Antibiotics destroy or halt the growth of micro-organisms. They do not harm body cells and can be used inside the body. However, microbes are becoming immune to many antibiotics so they don't work as well.
- **Vaccines** have been produced to prevent a person getting some bacterial and viral diseases.
- **Viruses** can be prevented by avoiding close contact with infected individuals.
- **Parasites** can be prevented by avoiding eating them or being bitten by their vectors. This often involves physical barriers like mosquito nets, washing hands or chemicals like insecticides.

Fighting disease

Using an assortment of special defences, the body normally manages to keep invading pathogens in check.

- **Non-specific resistance** — these respond in the same way to all invading organisms. They provide a variety of physical and chemical barriers against infection. Sweat, tears and saliva wash pathogens away and they also contain an enzyme that attacks them. The body also produces special white cells that engulf invading foreign cells.

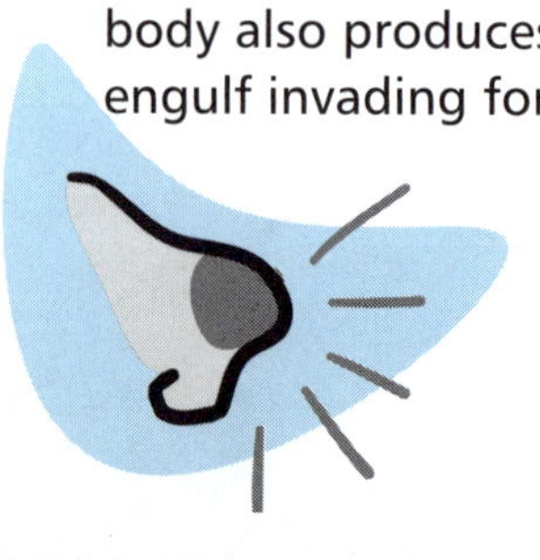

- **The immune system** — the immune system produces substances that are specifically selected to attack a particular invading pathogen. The immune system recognises specific foreign cells when they enter the body, and produces special chemicals called antibodies to fight them. The immune system also 'remembers' invading pathogens and responds even faster if they reappear.
- **Antibodies** are special proteins that are produced by cells of the immune system and circulate in the blood. When the pathogen enters the body, antibodies lock onto the foreign substances, or **antigens** on the pathogens surface. This immune response makes the invader harmless. Antibodies are specific — an antibody will react only with the antigen that triggered its formation.

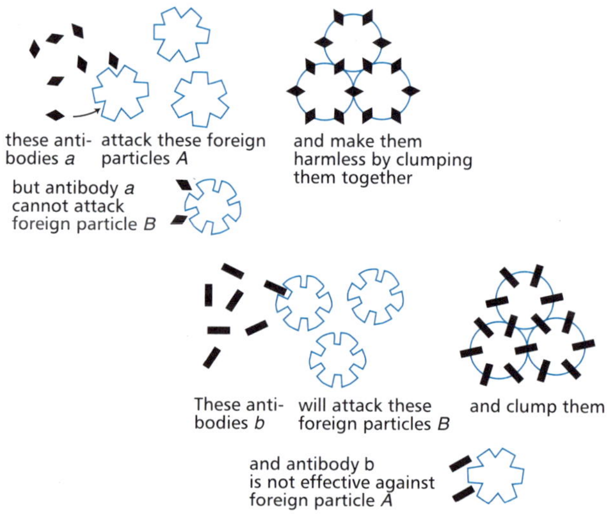

Antibodies are specific

Antibodies are produced by special white blood cells called lymphocytes. **Lymphocytes** can make antibodies against millions of potential pathogens. Some white blood cells digest the microbes. This is called phagocytosis.

- **Immunity** is the result of an immune response. Once the body has been infected with a pathogen it 'remembers' the pathogens' antigens. If the pathogen enters the body again, the body quickly increases its production of antibodies, and stops the pathogen from causing illness.

- **Immunisation** is the injection into the body of some dead or weakened pathogens that trigger the body's immune system to produce the appropriate antibodies. The person does not then get these diseases. However some people do not agree with immunisation, as it can cause some side effects.
- **Autoimmune disease** — sometimes the immune system breaks down and the body starts to attack its own tissues. In rheumatoid arthritis the antibodies attack antigens in the joints, which makes the joints very painful.
- The **lymphatic system** drains fluid from the body's tissues into the blood. It also houses lymphocytes that fight infection. **Lymph nodes** occur at intervals along lymph vessels. They filter out bacteria and other foreign material. They often swell up if the body is fighting infection.

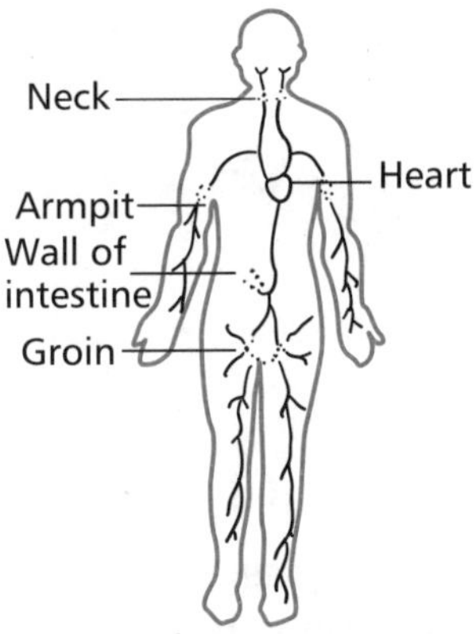

The lymphatic system and lymph nodes

- **Allergies** are the result of the immune system working too hard and harming the body as well. Allergies are triggered by substances called allergens, for example, hay fever is caused by contact with pollen. Most responses involve the release of histamine, which can make breathing difficult.

Glossary

Allergy
An excessive immune response to an antigen.

Antibodies
A protein that locks onto a specific foreign substance.

Antigen
Foreign substances on the surface of a pathogen.

Bacteria
Organisms made of one cell, they are neither animals nor plants.

Disease
A breakdown in the body's stable state.

Homeostasis
The maintenance of stable conditions in a cell or a living thing.

Host
An organism that provides food for a parasite.

Immune system
Specific defences against disease-causing organisms.

Immunity
Resistance to a pathogen or a foreign substance.

Lymphatic system
A system of channels that drain fluid and fight infections.

Lymph nodes
Swellings in the lymphatic system.

Lymphocyte
Special white blood cells that produce antibodies.

Non-specific resistance
Defence systems that attack a wide range of pathogens.

Parasite
An organism that benefits at the expense of the host.

Pathogen
A disease-producing organism.

Vector
A living thing that carries a parasite or pathogen from one host to another.

Virus
Disease-causing organisms smaller than bacteria, they can only grow inside a living cell.

Science skills

Question 1

What process is used to protect the following foods from bacterial infection?

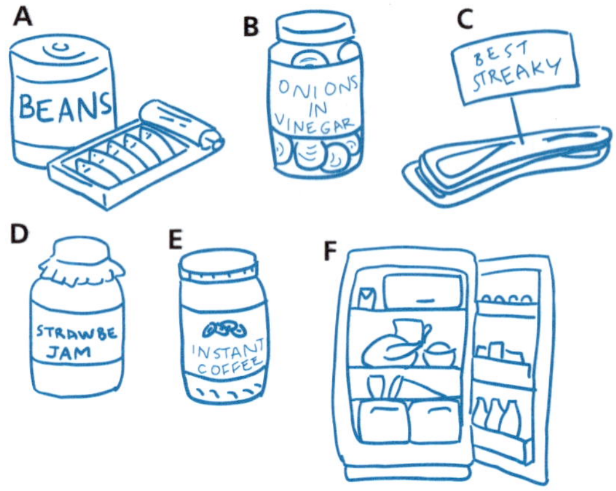

Question 2

Antibiotics are taken to kill bacteria. Study the graph below:

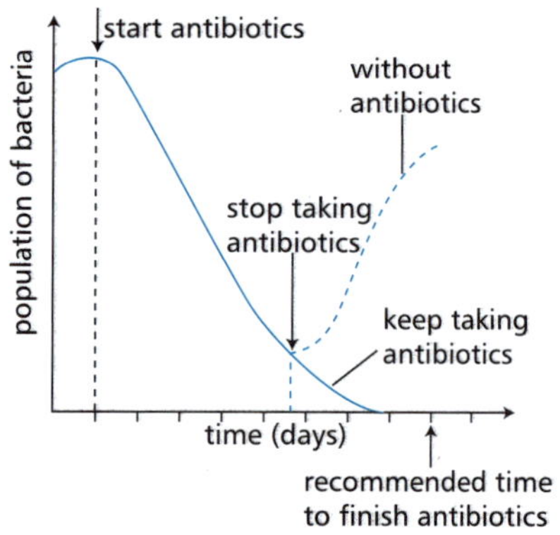

a How many days did it take for the antibiotics to kill the bacterial population?

____

b What happens if the patient does not take all the antibiotics prescribed?

____

____

Question 3

Some microbes are useful. Write down the jobs that microbes do in the following situations.

Useful microbe sites	What they do
compost heap	
bakery	
sewage treatment	
cow's stomach	

Question 4

The graph shows infection and recovery time for a person with influenza.

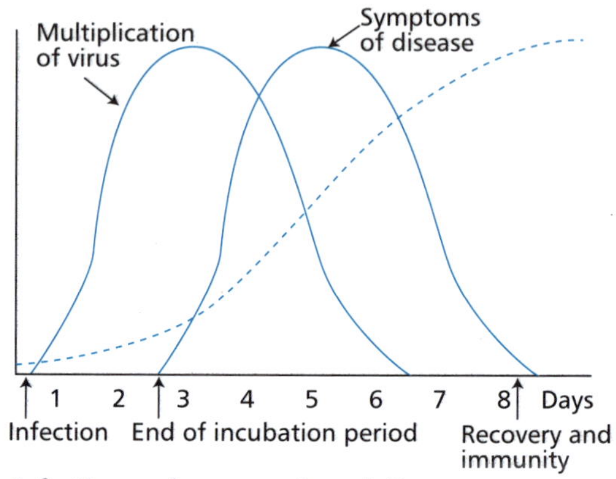

Infection and recovery from influenza

a How long is the incubation time for influenza?

____

b What happens in the body during this time?

____

c How long does the disease usually last?

____

d Why is it that other people can catch a virus from you before you even know you have it?

____

____

e What do you think the dotted line represents?

____

Question 5

Viruses	Bacteria

Write the following statements under the correct heading.

- Much smaller than a single cell.
- Measles, chicken pox, and mumps.
- Food poisoning, whooping cough and tonsillitis.
- About the size of a single cell.
- The body's immune system attacks and destroys it. Antibiotics are effective.
- The body automatically produces antibodies but antibiotics have no effect.
- Takes chemicals from cells — cannot exist on their own.
- Absorbs nutrients and divides and multiplies like normal cells.

Question 6

The following habits cause pathogens to be spread very quickly.

Complete the last word of each sign.

Question 7

According to the World Health Organisation, overuse of antibiotics, world travel and poverty have combined to produce an increased spread of infectious diseases.

Look at the graph below and answer the questions.

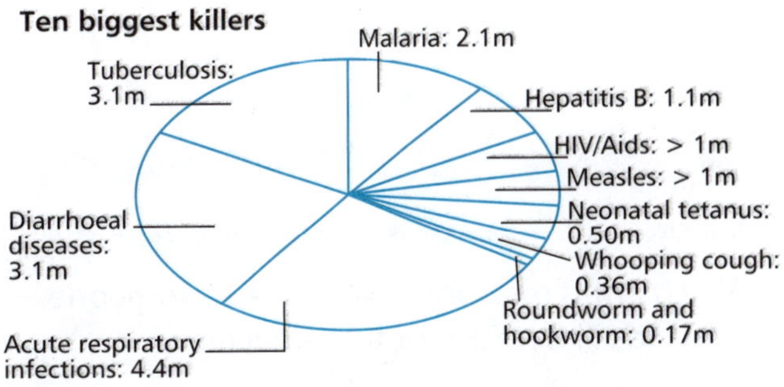

a Which disease causes the most deaths?

____

b Which one causes the least deaths?

____

c How many people does malaria kill per year?

____

Question 8

The flow chart below shows how saprophytic bacteria keep things moving around.

Fill in the boxes with the correct words from the following: living plants, simpler substances, dead animals and plants, living animals.

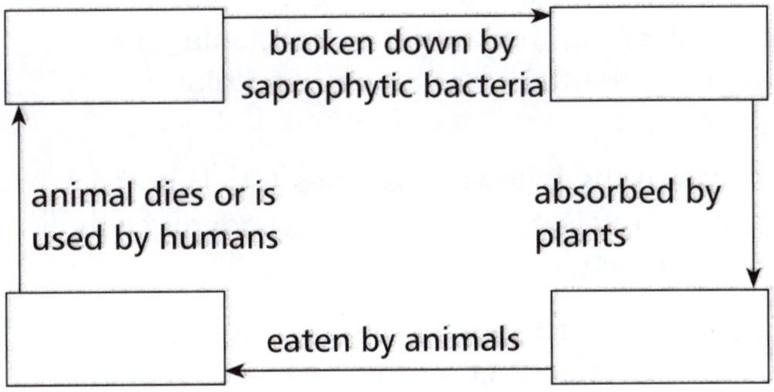

Question 9

Antibiotics can stop bacteria growing or kill them. Students soaked two paper discs in different antibiotics. They put them on agar plates in which bacteria were growing. Study the results.

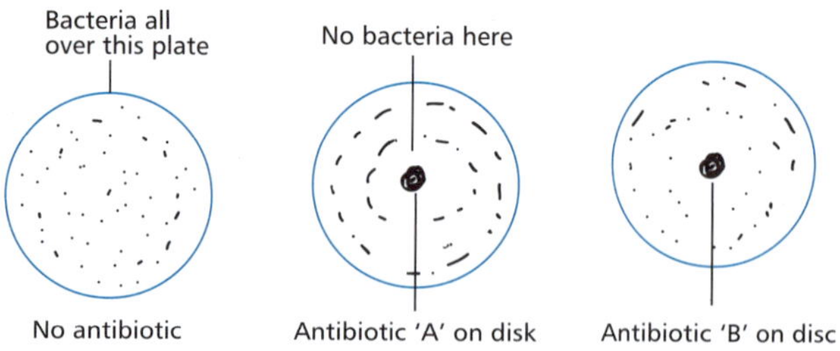

a Which antibiotic would you give to someone who was ill because of these bacteria?

____

b Why did you choose this one?

____

____

c Why did the students use the agar plate without a disc?

____

____

Question 10

A test for safe drinking water uses the fact that bacteria vary in their growth rates at various temperatures. If drinking water is held at 22 °C and counts are made, most of those that develop are harmless organisms. However if the water is held at 37 °C, the organisms developing are mainly potential parasites of warm-blooded animals.

Which of the following samples A to D in the table below is most likely to be condemned as drinking water.

Sample	Number of bacteria developing at 22 °C	Number of bacteria developing at 37 °C
A	high	medium
B	low	high
C	medium	low
D	low	low

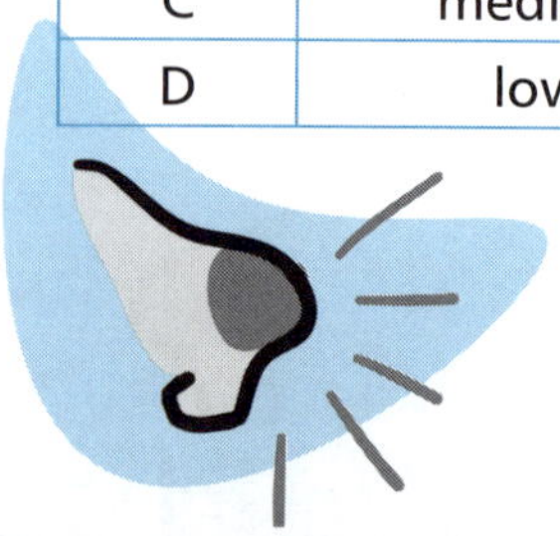

Revision questions

Question 1

a What causes influenza?

____

b Why is flu so easy to catch?

____

____

c How does sneezing spread flu?

____

____

Question 2

What type of microbe causes the following diseases?

Disease	Organism
athlete's foot	
chicken pox	
measles	
malaria	
salmonella poisoning	
tuberculosis	
AIDS	

Question 3

Write down three ways that pathogens can get into your body.

____

____

____

Question 4

Complete the sentences below with the correct words.

The immune system protects you from ______, ________ and other micro-organisms. When your body is invaded by a virus the ____________ produce __________ which disable the virus and prevents it from entering the body's cells.

Question 5

Match the following words with the correct explanation.

Sterile	A substance that stops the growth of disease-causing organisms in wounds.
Disinfectant	A substance that stops certain bacteria from reproducing.
Antiseptic	Something that has no viruses, bacteria or other organisms on it.
Antibiotic	Something that kills disease-forming organisms, e.g. heat and sunlight.

Question 6

a List two things that are the same about viruses and bacterial diseases.

____

____

____

____

b Are antibiotics useful in the treatment of influenza?

____

c What is the difference between an antibiotic and an antibody?

____

____

____

Question 7

Complete the table below with the symptoms of each disease and the method of transmission.

Disease	Symptoms	Transmission
ring worm		
chicken pox		
AIDS		
salmonella		
measles		

Question 8

a Some white blood cells make __________ to destroy microbes, others _________ them. White blood cells make new kinds of antibodies when you are infected by, or immunised against, a new disease. ___________ are chemicals that kill ___________ in your body.

b Which diagram below explains phagocytosis and which the destruction of invaders via antibody production?

A

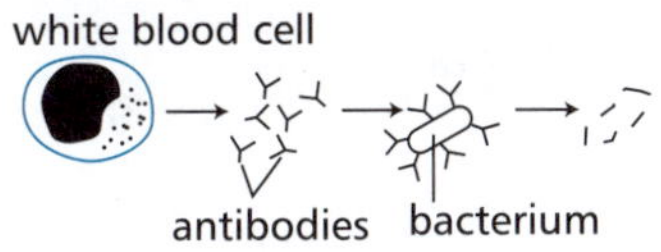

B

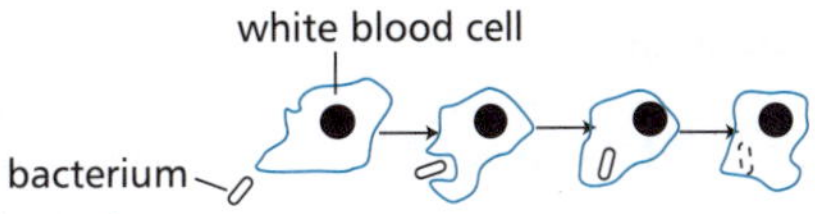

Question 9

Write explanations for the following terms.

Term	Description
vector	
virus	
antigen	
antibody	
parasite	
pathogen	

Question 10

Which of the following is not an infectious disease?

a AIDS

b whooping cough

c athlete's foot

d heart disease

Question 11

Viruses can best be described as:

a unusually small bacteria

b toxins causing disease

c minute infectious particles

d the cause of coughs and colds, and influenza

Question 12

Which of the following would not be seen as an effective way of controlling a viral infection?

a isolation

b antibiotics

c vaccination

d preventing its spread

Question 13

Give two diseases caused by each of the following pathogens.

Pathogen	Disease 1	Disease 2
fungus		
virus		
parasite		
bacterium		

Question 14

____________ is the injection into the body of some __________ or __________ pathogens that trigger the body's immune system to produce the appropriate ____________. The person does not get the disease.

Question 15

Match the following words with the correct explanation.

Immune system	Injection of dead or weakened pathogens and the person does not get the disease.
Immunisation	Once the body has been infected it 'remembers'.
Immunity	Produces substances specifically for attacking invading organisms.

Question 16

a What is an allergy?

____

____

____

b What substance is released during an allergic response?

____

c Give an example of an allergy.

____

Question 17

The ____________ ________ drains fluid from the body's tissues into the blood. It also houses ____________ that fight infection. ________ ________ occur at intervals along lymph vessels. They filter out bacteria and other foreign material. They often swell up if the body is fighting infection.

Question 18

The graph below illustrates an infection in a person. What happens at A, B, C and D?

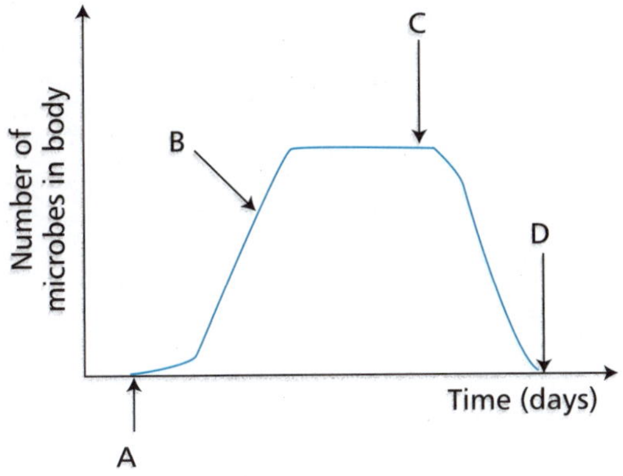

Question 20

a Which blood cells defend the body against disease?

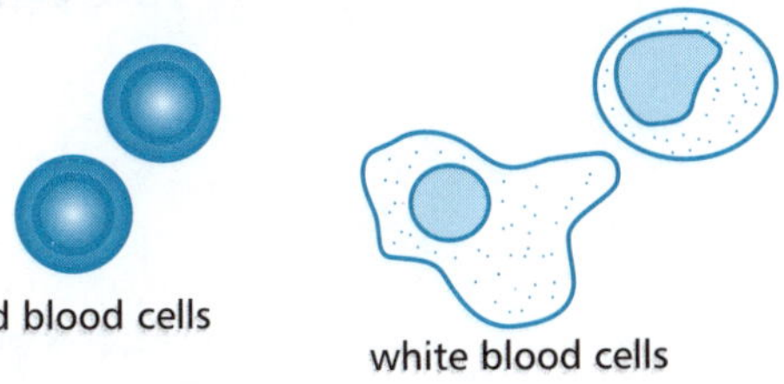

b How do they do this?

c Once you have had a disease do you normally get it again?

Question 19

In which section (1–6) are bacteria used to clean up sewage?

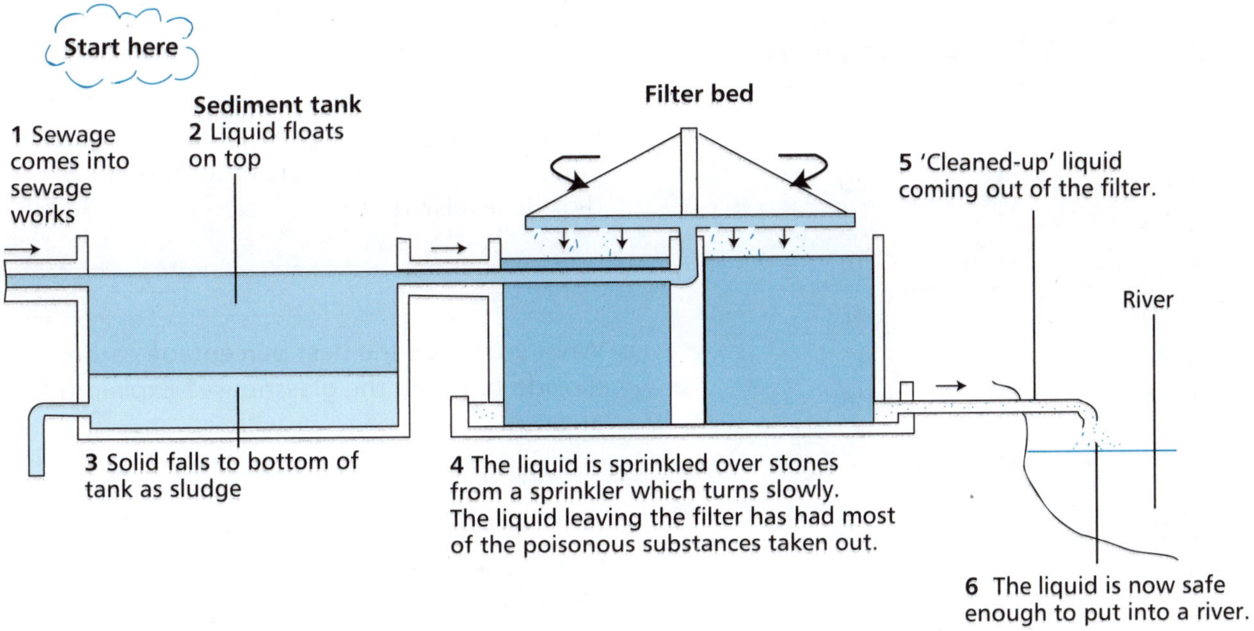

Test A

Question 1

Fill in the missing words.

Carbohydrates are packed with __________.

Animals use energy from carbohydrates such as __________ and __________ to fuel their cells.

Plants use a process called ______________ to make their own ______________.

Question 2

An experiment was conducted where plants were grown in different concentrations of carbon dioxide. Their rates of photosynthesis are recorded below.

Carbon dioxide concentration in air	Rate of photosynthesis
0	0
0.02	20
0.04	29
0.06	35
0.08	39
0.10	42
0.12	45
0.14	46
0.16	46
0.18	46
0.20	46

a Draw a graph of the results.

b If the carbon dioxide concentration in air was 0.05% what is the rate of photosynthesis?

____

c Why does the graph level off?

____

____

d If the plants could be grown in a glasshouse would it be possible to increase the carbon dioxide levels? How?

____

____

e What would be the best percentage carbon dioxide to use in the glasshouse? Explain this answer.

____

____

f Why would nurseries want their plants to grow fast?

____

____

Question 3

a Label the following diagram with the words:

mouth, oesophagus, stomach, small intestine, large intestine, pancreas, liver

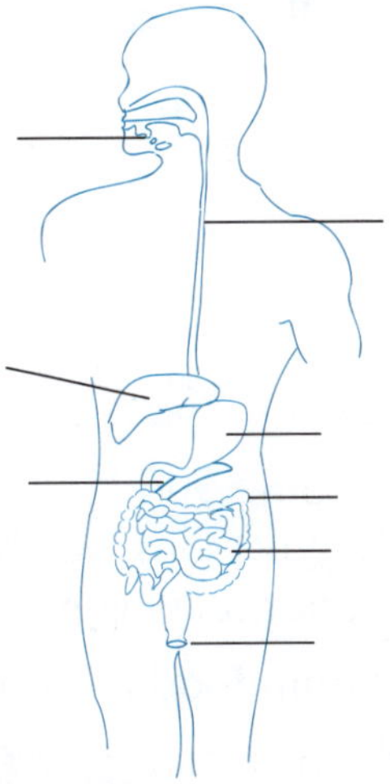

b Complete the table below by filling in the spaces.

Organ	What it does
	breaks up food and digests starch
oesophagus	
	juice (bile) splits up fats into tiny drops
stomach	
	digestion and absorption of sugars, proteins & fats
pancreas	
large intestine	
	expulsion of waste material

Question 4

Only soluble molecules can pass into your blood. The food you eat is a mixture of large molecules which must be digested into small molecules.

What are the names of the small molecules that the following foods are digested into?

a 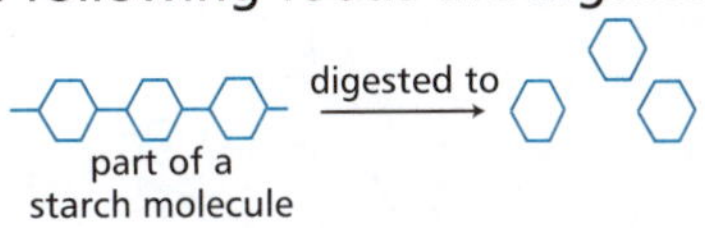

b 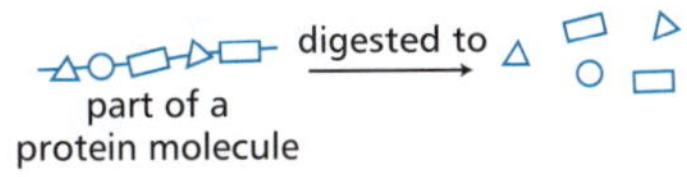

c 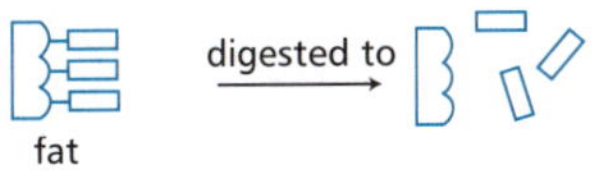

____

____

____

Question 5

a Put the following four sentences into the correct order so that they explain how food from your digestive system gets to your cells.

☐ Your blood transports small molecules around your body.

☐ You absorb the small molecules into your blood stream through the walls of the small intestine.

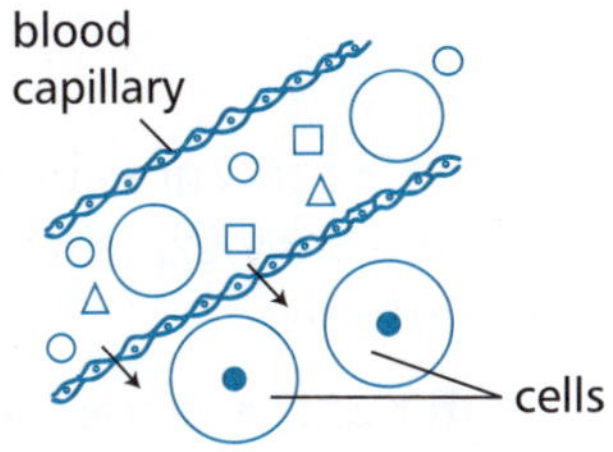

☐ Enzymes break food down into small molecules in your digestive system.

☐ They pass through the walls of the capillaries into your cells where they are used.

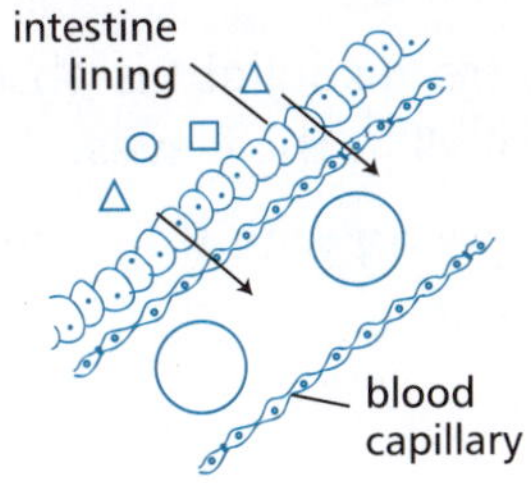

b Fill in the table below.

Respiring cells take in:	Respiring cells give out:

Question 6

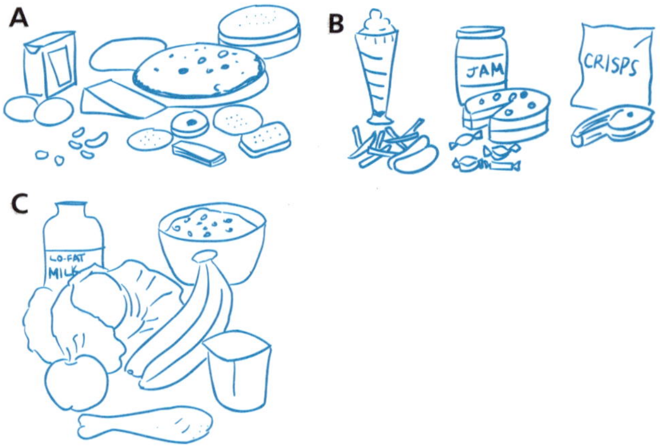

a To stay healthy, which group of food should you eat the most of?

____

b Which group should you eat only a little of each day?

____

c What do the foods in group B have high levels of?

____

Question 7

a Number the sentences below in order from 1 to 4 so that they explain the process of photosynthesis.

____ Water and minerals pass into the roots.

____ Oxygen passes into the air from the leaf.

____ Carbon dioxide passes into the leaf from the air.

____ Energy from sunlight is absorbed by the chlorophyll in the leaves.

b Label the drawing with the number of the correct sentence.

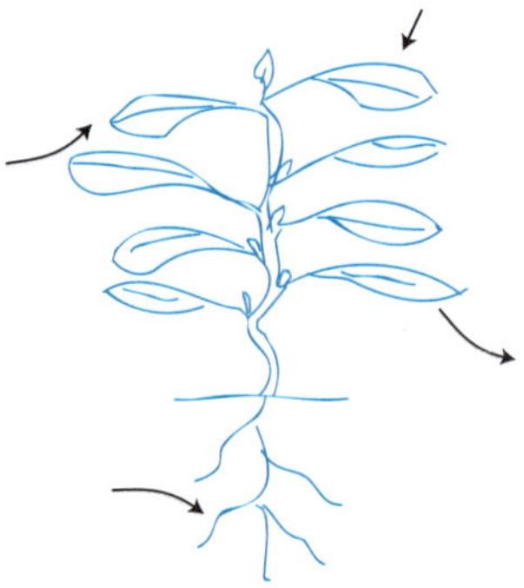

Question 8

Add the correct balance or the correct person to the sentences below:

Balances

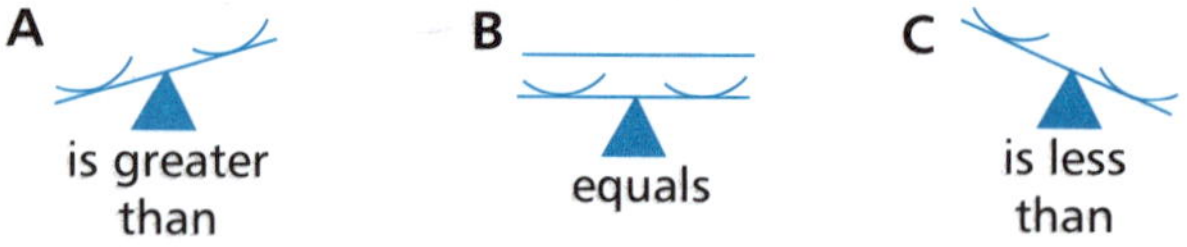

Persons

When the energy from a food (   ) the energy needs of the body, a persons weight remains the same. They do not put on weight or take it off.

When the energy from food (   ) the energy needs of the body the result is ________.

When the energy from food (   ) the energy needs of the body this is the result ________.

Question 9

a Label the parts of the eye.

A ____________

B ____________

C ____________

D ____________

E ____________

F ____________

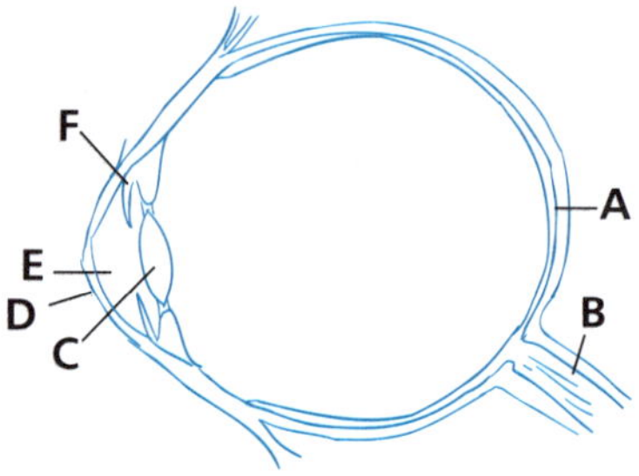

b Complete the sentences below that explain how we see.

Light passes through the c ____________, a____________ h____________, p____________, l____________ and v____________ h____________ to strike the retina. The ____________ focuses the light on a small section of the ____________. The ____________ stimulates the receptors, which send impulses through the ____________ nerve.

Question 10

a Label the diagram below with the correct areas of taste.

b The taste buds are clusters of cells that sense ____________. They are connected by ____________ to the brain. When you eat, substances in food trigger the cells to send ____________ to the brain. Taste together with ____________ produces the food's flavour.

Question 11

a Hearing enables animals to pick up sounds, communicate and to detect danger.

Label the diagram below with the following words:
external (auditory) canal, eardrum, semicircular canals, cochlea

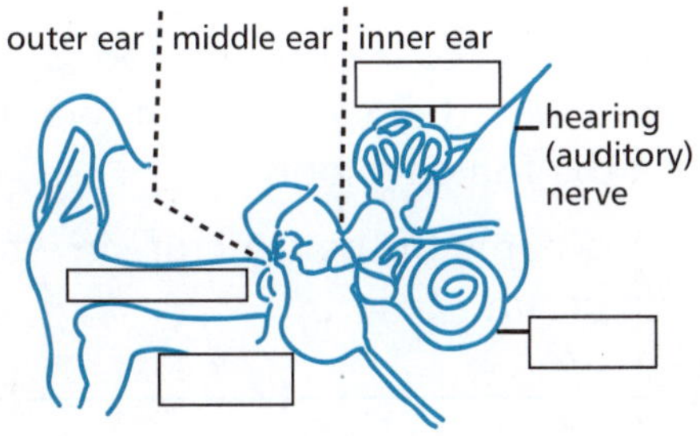

b Some animals use high frequency sound to build up a picture of their surroundings. Name two animals that use echolocation.

____

Question 12

The following cycle of events is called the menstrual cycle.

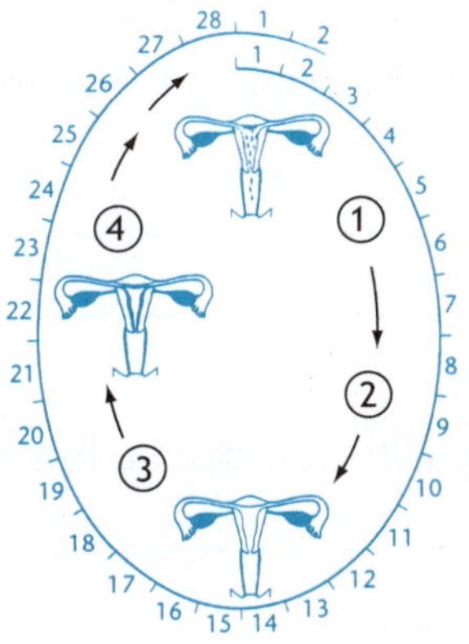

a How long does the cycle last?

____

b Why do women have a menstrual cycle?

____

c Match the following explanations at points 1 to 4 in the cycle.

Explanations:

- If the egg is not fertilised it dies.
- During the period of bleeding, the lining of the uterus is shed.
- The lining of the uterus continues to thicken.
- A new lining grows inside the uterus ready for a fertilised egg to grow. An egg is released from the uterus.

Question 13

The following table provides some information about survival rates in vertebrates.

Animal	Fertilisation	Egg	Survival rate (%)
amphibian	external in water	100s in jelly	15
fish	external in water	1 000s to 1 000 000s	5
reptiles	internal	1–50 leathery shell	70
birds	internal	1–10 chalky shell	50
mammals	internal	1–10 develop inside female	Less than 95

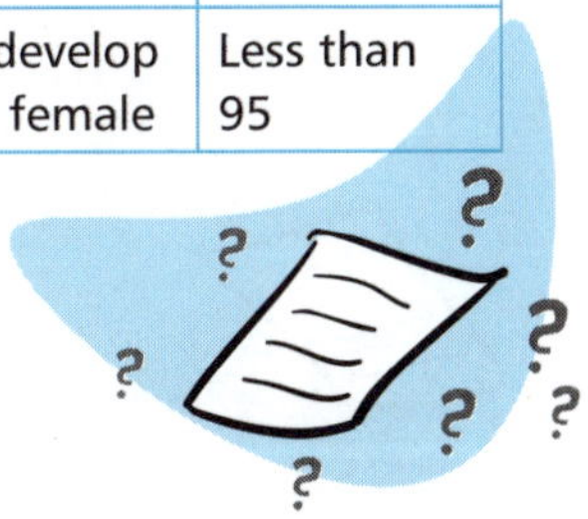

a Explain the difference between internal and external fertilisation.

b What type of fertilisation is common in aquatic animals? Why?

c Explain why fish have such a low survival rate.

d Why do birds need a solid eggshell?

e Is this the same for reptiles?

f Why do mammal offspring have such a high survival rate?

Question 14

Label the human female sex organs in the diagram below.

FRONT VIEW

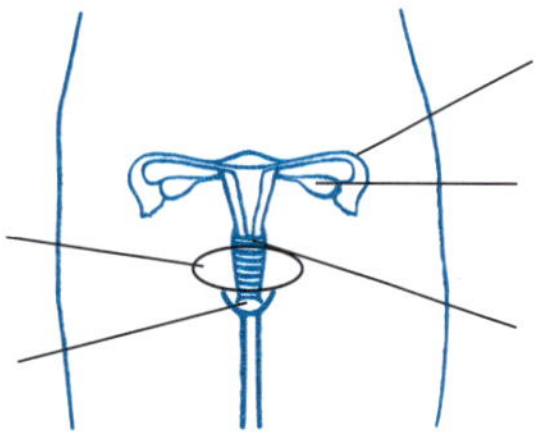

SIDE VIEW

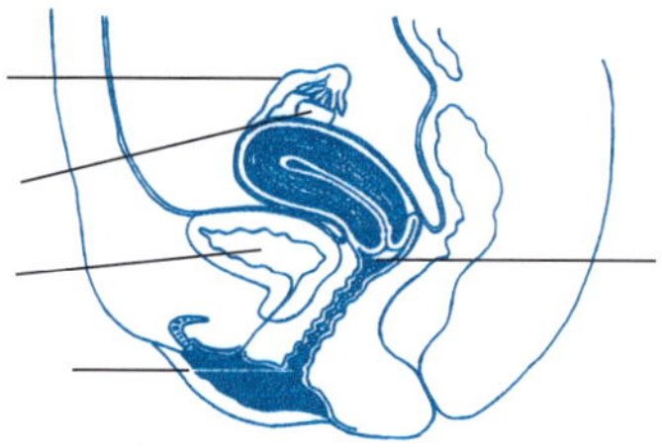

Question 15

a Look at the diagram below and label the two cells.

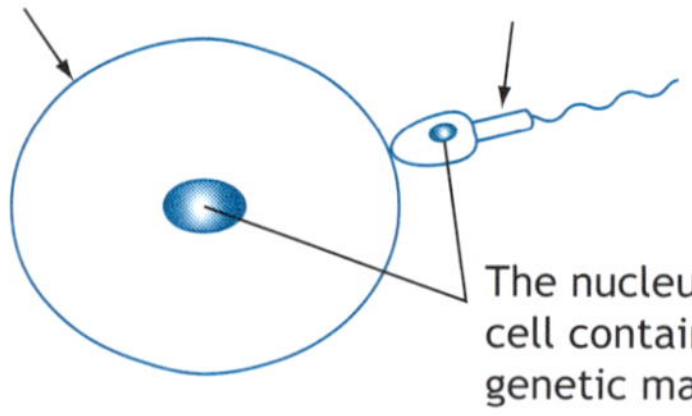

b In brackets write the words haploid or diploid where appropriate.

c What is this process called?

d What is the fertilised egg called?

e Is the fertilized egg a haploid or diploid cell?

Question 16

The diagram shows the feeding of a foetus.

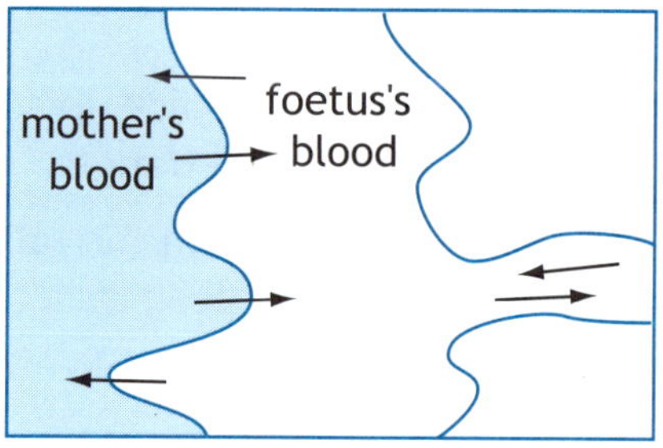

a Label the arrows to show two products that pass into the mother's blood and two that pass into the foetus' blood.

b What is the name of the structure in which this exchange occurs?

c What is the name of the structure that carries blood to and from the foetus?

Question 17

Are the following examples (a) and (b) of asexual or sexual reproduction? Explain your answer.

A **BUDDING**

B **DIVISION**

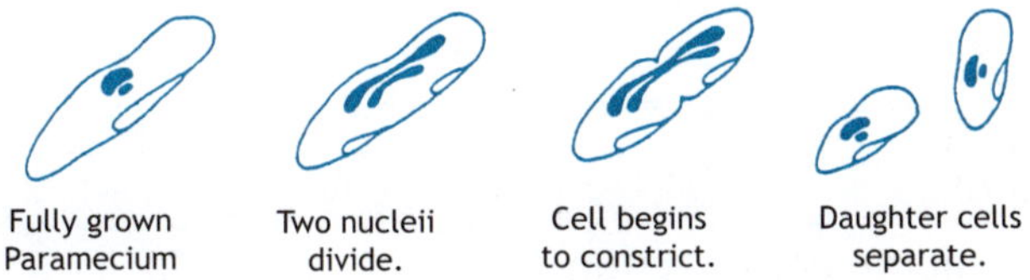

Question 18

Match the following stages of development with the correct explanation.

Foetus	A single cell resulting from the union of a sperm and an ovum.
Zygote	All developmental stages from the ninth week to the birth.
Embryo	A ball of cells with surface layer and inner cell mass.
Blastocyst	All development stages from two weeks until the end of the eighth week.

Question 19

Complete the table below with the function of the male reproductive structures.

Structure	**Function**
testes	
epididymis	
vas deferens	
seminal vesicles	
prostate gland	
penis	
seminiferous tubules	

Question 20

The human female cycle is controlled by hormones. The ovaries make two hormones — oestrogen and progesterone.

a Label the graphs of the hormones in the diagram below.

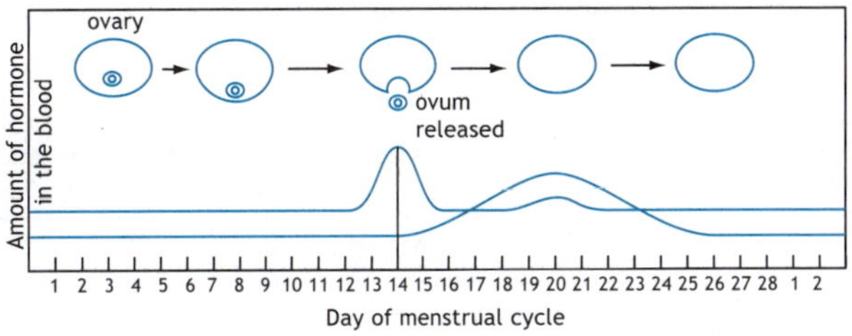

b Which hormone makes the ovary produce an ovum?

____

c What hormone does the testis of the human male produce?

____

Question 21

Scientists are concerned about the number of antibiotic resistant strains of bacteria that are appearing.

a How would a lack of antibiotics to treat diseases affect you?

____

____

____

b Why do they think this is happening?

____

____

c What can we do to reduce this problem?

____

____

____

____

Question 22

Botulism is very bad form of food poisoning.

a How do you get food poisoning?

____

b How can you prevent food poisoning? What is the name of another form of food poisoning?

____

____

c Study the following table and decide which foods are likely to be responsible.

Person	Appearance of symptoms	Food on menu: Soup	Prawn cocktail	Chicken salad	Sea food	Spaghetti sauce	Apple pie & cream	Fruit salad & ice cream	Pavlova
A	Yes			✓			✓		
B	Yes	✓		✓			✓		
C	Yes			✓				✓	
D	Yes	✓		✓			✓		
E	Yes			✓			✓		
F	NO SYMPTOMS		✓		✓				✓
G	NO SYMPTOMS				✓		✓		
H	NO SYMPTOMS	✓			✓				✓
I	NO SYMPTOMS					✓		✓	
J	NO SYMPTOMS	✓			✓		✓		
K	NO SYMPTOMS	✓		✓				✓	
L	NO SYMPTOMS		✓			✓		✓	

d If more than one item is suspected, how could contamination have occurred?

Question 23

a Which diagram is a virus and which is a bacterium?

b Which is bigger, the virus or the bacterium?

A

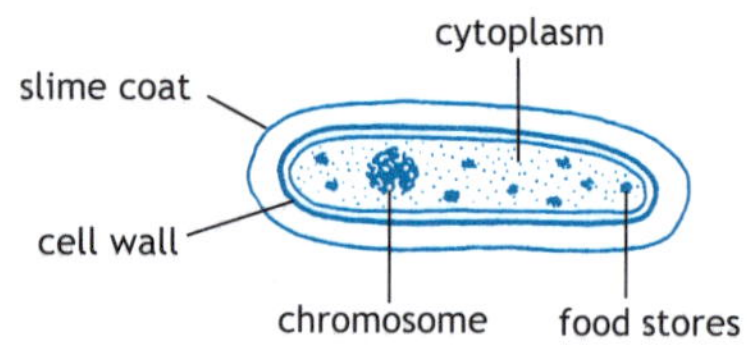

B

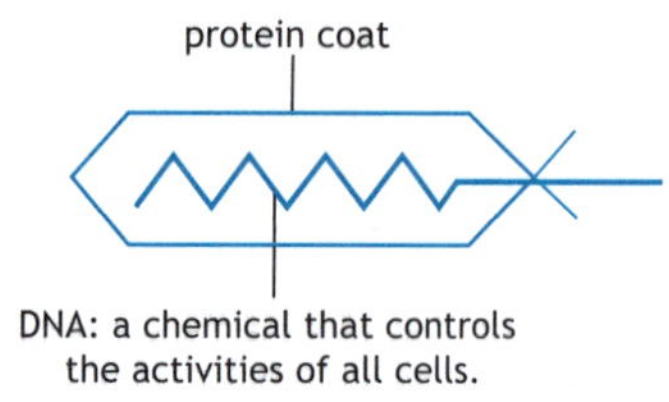

Question 24

The following table lists some diseases. Mark whether they are caused by viruses or bacteria.

Disease	Virus	Bacterium
warts		
pimples		
mumps		
diarrhoea		
typhus		
AIDS		
middle ear infection		
influenza		

Question 25

a What is a disease?

b What cause diseases?

c List the barriers that the body has to resist infection.

CHAPTER 5

CHEMICAL REACTIONS

How fast can reactions go?

When something changes we say that it is different. If the change makes a new substance then it is a called a **chemical change**. If the change does not make a new substance, then it is a physical change. Chemicals can meet without reacting, but if a new substance is formed from the existing ones, then a chemical reaction must have taken place. There may be other easily seen evidence of a chemical reaction such as:

- the production of a gas
- a colour change
- a temperature change

Chemical changes always make new substances and can rarely be reversed. However, physical changes, like changing ice to water, are usually easy to reverse.

Chemical reactions are taking place in the world around us all the time. Living things actually depend on chemical changes.

Chemical reactions and energy

Energy makes things happen. Every chemical reaction has energy changes. Sometimes these energy changes are the most important part of the reaction — petrol for powering cars and chocolate cake for powering people! All chemical reactions involve energy.

A candle burning is an example of a chemical reaction — candle wax reacts with oxygen in the air to form water vapour and carbon dioxide. Heat energy and light energy are produced. The match supplies the heat energy to make the wax hot enough to combine with the oxygen. After that the heat energy from the reaction is enough to keep it going.

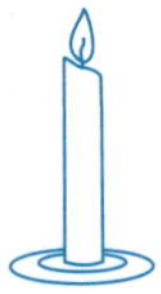

Types of chemical reactions

There are many different types of chemical reaction.

Oxidation reactions

Oxidation occurs when a substance combines with oxygen. Some oxidation reactions create big problems and others are very useful.

- Burning or **combustion** is a very common chemical reaction. When a substance burns it combines with oxygen in the air, producing heat, water, carbon and carbon dioxide. The burned substance is called a fuel.

  Different elements burn with different coloured flames — for example boron (green), sodium (orange), potassium (mauve).

  A motor car uses an internal combustion engine which burns petrol to provide the power to keep the car going.

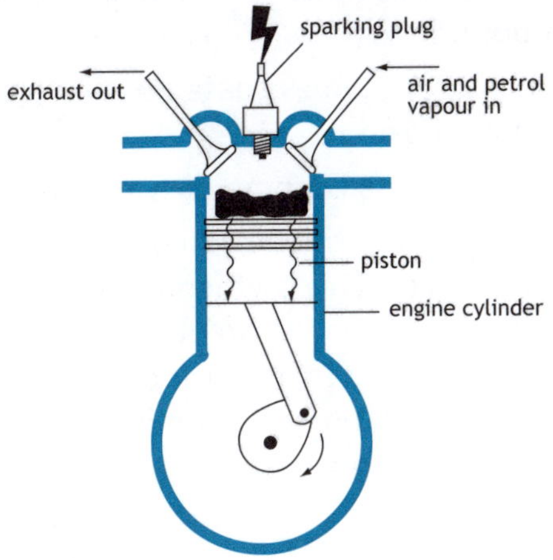

Engine cylinder

The petrol vapour and air is sucked in from the carburettor and compressed. As it is compressed it gets hotter and then a spark from the spark plug ignites the mixture and drives the piston down. Waste gases and heat are expelled through the exhaust system.

- Oxidation reactions do not always involve burning. **Corrosion** is the result of metals reacting with air, water or other substances in their surroundings. Iron and steel corrode in damp air to form rust (iron oxide), which is a different substance from the original metals. If a metal is very reactive it will burn brighter and react faster with substances such as oxygen.

  Most methods of slowing rusting down put some kind of non-reactive barrier between the iron and the air and water such as:

  - turning iron into stainless steel (stainless steel sinks). This material is more expensive.
  - painting the iron (garden furniture).
  - coating the iron with oil or grease (bike chains and car engines).
  - coating the iron with plastic (dish racks, wire).
  - coating with other metals, like zinc (roofs, garden sheds) and tin (steel food cans are coated in tin) which are not as reactive as iron. This process is called galvanising.

  The less reactive a metal is, the less likely it is to corrode.

  Other metals also corrode:

  - Copper can become coated with a greenish compound.
  - Aluminium becomes coated with white aluminium oxide
  - Silver and brass become dirty or tarnished.

- Living things use oxidation to break down food substances. This is when the cells combine the oxygen we breath with the carbon and hydrogen in our food to release energy. This oxidation reaction is called **aerobic respiration**.

The equation for aerobic respiration follows:

glucose + oxygen → carbon dioxide + water + energy

Reactions with liquids

- **Water** — Some metals like magnesium react slowly with water but more quickly in water vapour (steam). Others like sodium react very quickly with water. Most of these reactions make a gas called hydrogen. Some metals, for example copper, don't react much at all with water.

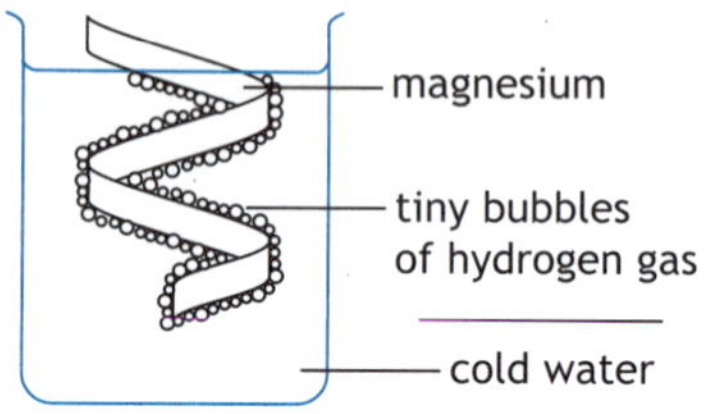

A. Magnesium in water

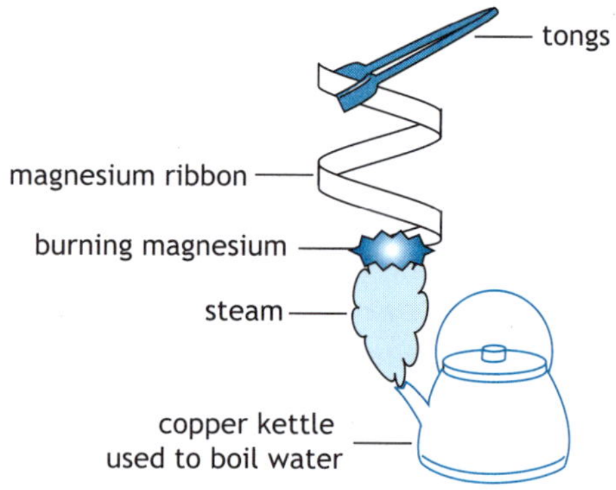

B. Magnesium in water vapour (steam)

Sodium is a very reactive metal. It reacts so quickly with water that it must be stored in oil to prevent it reacting with water vapour.

The word equation for this reaction follows:

sodium + water → sodium hydroxide + hydrogen

- **Acids** — Vinegar and lemon juice are weak acids and are harmless. Many drinks have weak acids in them as people like the tangy taste. However others such as sulphuric acid are very strong. Sulphuric acid is very corrosive and dangerous to handle. Strong acids are very reactive but weak ones are not.

  Substances react with acids. For example, if we add magnesium to dilute hydrochloric acid it reacts to produce new substances — magnesium chloride and hydrogen gas.

Acids are usually clear, colourless liquids. When they are dissolved in water we cannot see them and we do not know how strong they are. Substances that behave in the opposite manner to acids are called bases and those that are neither an acid nor a base are called **neutral**. Some bases dissolve in water and are called alkalis (e.g. NaOH). Strong alkalis can also be very corrosive.

We can use special dyes called **indicators** to check for acids. Universal indicator changes colour according to how acidic or alkaline a solution is.

We use a special scale to measure acidity. It is called the **pH scale**.

red	orange	yellow	green	blue	navy blue	purple
0–2	3–4	5–6	7	8–9	10–12	13–14

← more acidic — neutral — more alkaline →

pH scale

Acids and bases react together to produce salt and water. However if you do not use enough base you get some acid left over as well. If you add too much alkali you get some base left over. If you mix just the right amounts, nothing is left over and the solution is neutral. We can use the indicators to work out exactly how much acid we should add to neutralise a quantity of base. Following are examples of the use of acids and alkalis in our daily life.

- Acid from food attacks your teeth so toothpastes are usually slightly alkaline.
- If you eat too much, too much acid is produced in your stomach, causing indigestion. Indigestion tablets are alkaline and they neutralise the acid. They are made of baking powder and give off carbon dioxide, which causes us to burp.
- Cake recipes contain baking powder and a weak acid. When these ingredients get wet (milk and eggs) they react and produce carbon dioxide that makes the cake rise.

Chemical reactions in living things

Every living thing needs chemical changes to stay alive. Our cells are the chemical factories where chemicals are manufactured and reactions occur. Cells are constantly breaking down substances and building up new ones. This two way process is called metabolism and includes all the living processes that take place in a living organism. Examples include:

- **Growth and Development** — for a fully formed bird to hatch from an egg, the yolk and the white of the egg must undergo many chemical reactions. The new substances formed are feathers, bones, blood etc.
- **Digestion** — Once an organism has ingested food it must be broken down by a series of chemical reactions called digestion.

Reaction time

Substances are made of particles. Particles from different substances must collide for a reaction to occur. If there are lots of collisions then the reaction will be fast. If they are not colliding much then the reaction occurs slowly. Natural reactions usually happen slowly. Such reactions include rusting (oxidation) and the dissolving of limestone by acid rain (rain and carbon dioxide combine to form a weak acid).

Things that make reactions occur at different speeds:

- The amount of the substance present — this is called the concentration of a substance. The higher the **concentration** the more particles there are to collide with.
- **Surface area** — if you chop wood into small pieces then it burns faster as there is more area available for oxidation reactions.
- **Temperature** — the faster the particles are going the more often they will collide.
- **Mixing** — this brings more of the reacting particles in contact with each other.
- **Catalysts** — these are substances that speed up reactions by 'collecting' the reacting particles, bringing them together and then moving away. Enzymes are biological catalysts — proteins manufactured by the cells that speed up chemical reactions in living things.

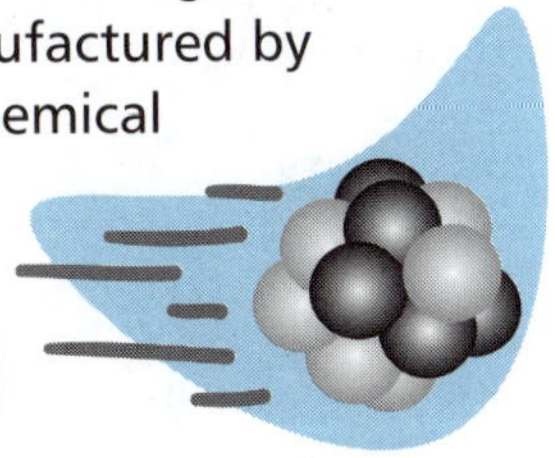

Glossary

Acid
A chemical compound containing hydrogen. Acids are very reactive, because in water the hydrogen atoms lose their single electron and become ions. Strong acids are very corrosive. Acids have a pH of less than 7.

Alkali
A substance that dissolves in water and neutralises acids.

Base
The opposite of an acid. Bases react with acids to form salts. Bases have a pH of more than 7.

Catalyst
A substance that speeds up a reaction without being used up.

Chemical Reaction
A process in which one or more chemical elements or compounds form new compounds.

Combustion
Burning — the rapid oxidation of a fuel.

Indicator
A chemical that changes colour depending on the concentration of hydrogen ions.

Metabolism
All the chemical processes that take place in a living thing.

Neutral
Neither an acid or an alkali (pH =7).

Oxidation
A large group of chemical reactions where a substance combines with oxygen.

pH Scale
A scale which is a convenient way to represent acidity in solutions.

Respiration
A chemical process in cells where glucose reacts with oxygen to release energy.

Science skills

Question 1

A family decides to have a barbeque. One person chops the wood into small pieces. Another person puts the pieces of wood into the barbeque and starts the fire with a match. After burning, the match and the wood are black.

a Which change to the wood makes new substances?

____

b Which is a chemical change?

____

____

Question 2

Burning is a chemical reaction. The elements in the burning substance join with oxygen from the air.

Number in order from 1 to 4, the following steps which describe the operation of the internal combustion engine.

___ The piston moves up to squeeze the mixture of petrol and air which gets hotter.

___ The exhaust valve opens and the waste gases and heat are expelled.

___ A spark from the spark plug ignites the mixture of air and petrol vapour.

___ A mixture of petrol vapour and air is sucked in from the carburettor.

Question 3

A student was interested in why things rust. Using iron nails she set up the following experiment and watched what happened to them over the following week.

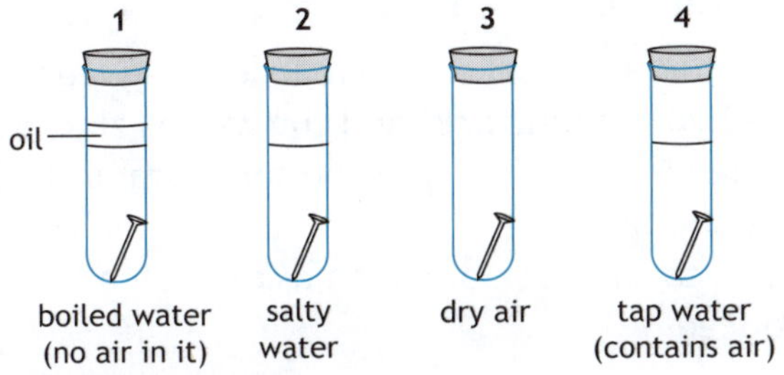

The results are tabled below:

Test tube	No rust	Rusty	Very rusty
1	*		
2			*
3	*		
4		*	

a Which nails did not rust?

b Why do you think the student put oil on top of the boiled water?

c Which nail rusted the most?

d What two substances are needed for iron to rust?

e Cars seem to rust more near the sea. Does the data support this idea?

Question 4

A sculptor makes two identical statues of brass. One is placed in Sydney and the other in Alice Springs.

a After 10 years which one will be the most corroded?

b Explain your answer.

Question 5

A teacher fills three jars with oxygen. He then heats up pieces of three different metals and puts them into the jars.

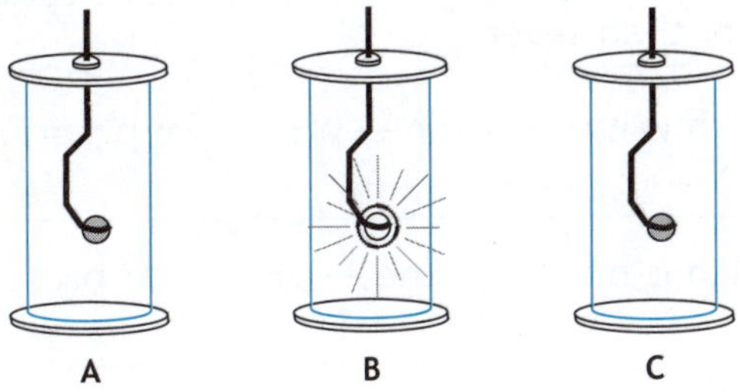

a Which metal burns the brightest?

b Which metal is the most reactive?

Question 6

Study Diagram 3 in the text.

a How can you see that a reaction is taking place in experiment (A)?

b What gas is made in both (A) and (B)?

c Why do you think the reaction is faster in (A) than (B)?

d Which substance is more reactive, the copper in the kettle or the magnesium ribbon?

e Why do we use copper pipes and not magnesium to carry water into our houses?

Question 7

Study the acidity or pH scale below.

The pH scale

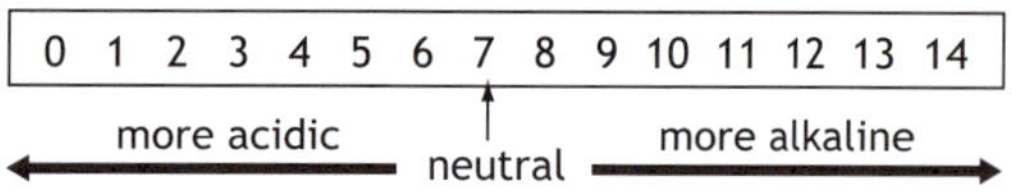

a Match the pH with the following substances — alkaline, neutral, acidic

pH	Substance
Equal to 7	
Less than 7	
Greater than seven	

b Which is more acidic — pH = 3 or pH =5?

c Which is most alkaline — pH = 8 or pH = 13?

Question 8

Universal indicator changes colour according to how acidic or alkaline a substance is. Using the table below, number substances 1 to 5 putting the most acidic first.

liquid	Universal indicator colour
A	Dark blue
B	Red
C	Yellow
D	Green
E	Orange

Question 9

Most cleaning products are either acids or bases. Students tested some household cleaners using universal indicator and recorded their results in the following table.

Product	Colour	Acid	Base
toothpaste	bluish		
bleach	dark blue		
shampoo	pale yellow		
dish detergent	yellowy green		
lemon juice	orange		

a Mark the product as an acid or an alkali.

b What makes acids and alkalis good cleaning agents?

Question 10

Some students studied the reaction between dilute hydrochloric acid and the metals zinc, iron and magnesium. They put the same amount of metal and acid in each flask and attached a balloon to the neck of each flask to catch any gas produced.

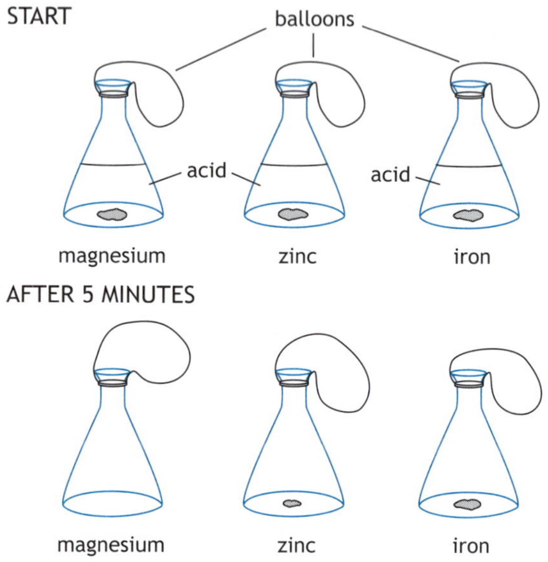

a Which produced the most gas?

b Which metal reacted the fastest?

c Which metal reacted the slowest?

d Write down the metals in order of their reactivity, with the most reactive first.

Question 11

A group of students were given a task designed to study reaction times. They were provided with Alka-Seltzer tablets and had access to any equipment they needed. They designed their experiment with constant volumes of 50 mL and a reaction time of 2 minutes. The reaction that produced the most gas (collected in a balloon) was considered to be the fastest.

Results are recorded below:

Table 1

Water temperature	Fastest reaction
Hot	*
Cold	

Table 2

Concentration	Fastest reaction
1 tablet	
2 tablets	*

Table 3

Surface area	Fastest reaction
Tablet not ground up	
Tablet ground up	*

Table 4

Mixing	Fastest reaction
Yes	*
No	

a What are the variables in this experiment?

b What are the constants?

c What combined process would cause the fastest reaction?

Question 12

A student carried out the following experiment. He quickly put 2 tablespoons of baking soda and 100 mL of vinegar into a bottle and pushed a cork into the opening.

a Predict what happened.

b What two substances reacted?

c What new substance was produced?

Question 13

The following experiment was set up to investigate the differences between iron and galvanised nails.

a Predict which nails will rust in water — the galvanised or the iron.

b Why is rust not useful?

c What will happen to the iron nail in the oil?

d How can we stop rust?

Question 14

The chart below shows some of the colours that are given when using universal indicator.

Colour					orange	yellow		green			dark blue				
pH	0	1	2		4		6	7	8	9	10	11	12	13	

a Fill in the missing colours and pH numbers.

b Mark the acid and alkali ranges.

Revision questions

Question 1

Burning is a chemical reaction. The elements in a burning substance join with oxygen from the air.

a What is burned in a car engine?

____

b Where does the reaction take place?

____

c What else is needed for this to happen?

____

Question 2

Fill in the blank spaces with the correct words.

The _______ used in most cars is _________, which is obtained from crude oil. When it undergoes _______________ (i.e. burning in oxygen) it provides the ______ needed to keep a car going.

Question 3

Why is painting the Sydney Harbour Bridge a never-ending job?

a It must always look nice.

b To keep the salty air off the metal and stop it corroding.

c So the metal is not too shiny in the sunlight.

d To keep people employed

Question 4

Predict what would happen to copper screws in a steel boat in salt water.

____

____

Question 5

Why do car exhausts rust so quickly? What are the advantages and disadvantages of using stainless steel exhausts?

____

____

____

Question 6

Consider the following metals: aluminium, copper, silver, iron, zinc, steel and tin.

Which ones will corrode the fastest? Why?

____

____

Question 7

a What is the process of galvanising?

____

b How does this prevent iron from rusting?

____

____

Question 8

Sodium is a soft metal that floats on water.

a Why is sodium stored in oil?

____

____

b What new products are produced by its reaction with water?

____

c Put the following metals in order of reactivity, starting with the most reactive.

copper magnesium sodium

____

Question 9

Students set up the following experiment:

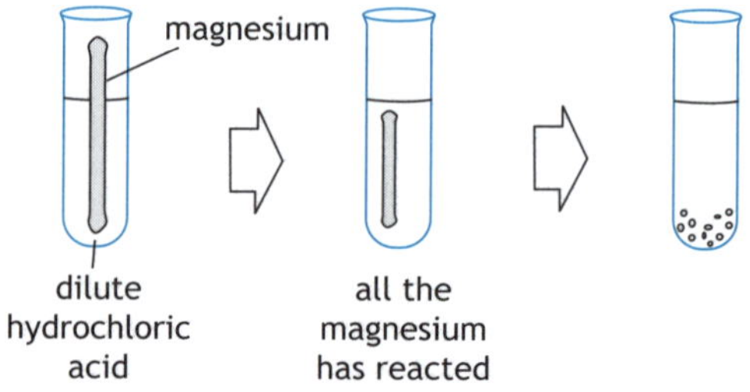

a What were the two original substances?

___

b What two new substances does this reaction make?

___

c Is this a chemical reaction or a physical reaction?

___

Question 10

Fill the spaces with the correct words:

a Chemical changes always make _____ substances. Life depends on ___________ changes.

b Some acids like ___________ acid are dangerous because they are very ________. Two acids that are not dangerous are ________ and ______ ______.

Question 11

How do barrier creams protect the skin from corrosive chemicals?

___

___

Question 12

When a match burns, a chemical reaction takes place. Look at the diagram:

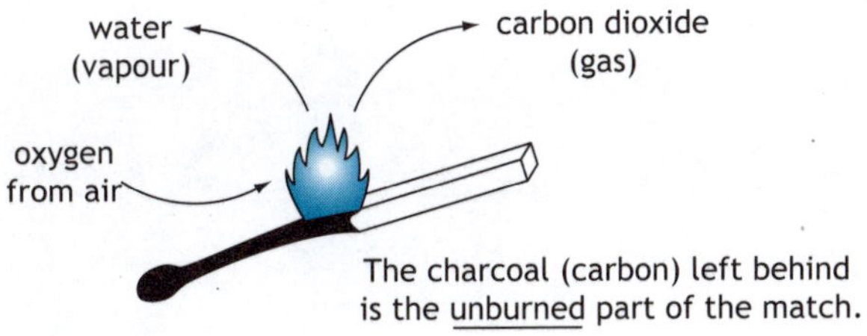

a What does the wood react with when it burns?

___

b What new substances are formed during the combustion of the match?

___

Question 13

Bees inject an acid when they sting you and wasps inject an alkali. What substances would be best used to relieve a bee sting and a wasp sting?

___

___

Question 14

Many of the limestone buildings and head stones in Europe are being eaten away by acid rain. What is acid rain?

___

___

___

___

Question 15

Galvanised nails are often used to keep roofing in place.

a What are the two main metals used to make these roofing nails?

___

b Why are the nails coated with a second metal?

___

___

Question 16

If the lining on cans is damaged why do the food contents spoil?

___

___

Question 17

Explain what the barrier is in each of the following pictures.

oiled bike chain

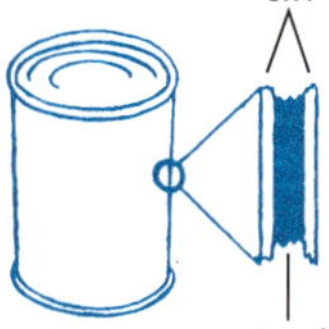

tin can

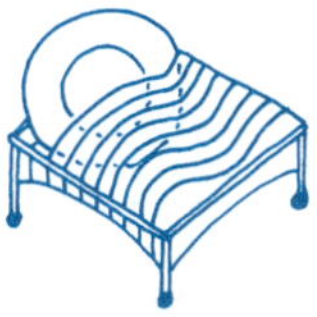

plastic-coated dish drainer

painted garden furniture

Question 18

Complete the sentences below with the correct words.

Rusting is a chemical reaction called ________. Rusting starts when iron or steel is in contact with ______ and ______. Rusting can be slowed dawn by providing a ________ between the metal and the atmosphere. The less reactive a metal is the less it is likely to ________.

Question 19

a How can we tell if something is an acid?

____

b When we mix an acid and an alkali we get

______________.

c Neutralisation reactions produce _______ and _______.

d Give an example of a useful neutralisation reaction.

____

Question 20

The bacteria that live in our mouths feed on sweet material caught in our teeth and produce acid. This acid reacts with the enamel covering the teeth and painful holes develop. The graph below shows how the pH in a person's mouth changes during the day.

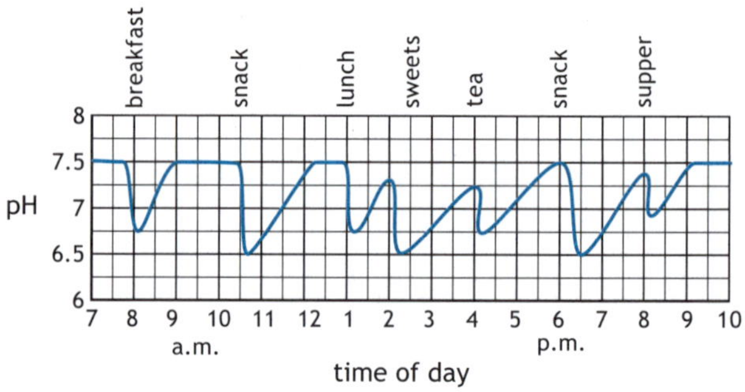

a When a solution gets more acidic does the pH get lower or higher?

____

b Why does the pH in a person's mouth get lower after each meal?

____

c What damage could be caused by the pH being low?

____

d Why is it a good idea to brush with toothpaste after each meal?

____

CHAPTER 6

THE UNIVERSE

Is anybody out there?

Scientists refer to the beginning of the universe as the **Big Bang** because evidence indicates that the universe exploded violently into being and has been expanding ever since. Apart from the galaxies near us, all the other ones are rushing away from us. The further away a **galaxy** is, the faster it and our galaxy are moving apart.

Somewhere between 13 and 20 million years ago all the matter in the universe was concentrated into a single point. Physicists believe that time is connected to the existence of space, matter and gravity. Before the Big Bang there was no time or space, so there was no 'before'. The explosion that expanded the point of matter at an unimaginable speed, created the universe. At the same time as this expansion, 'time' was created as well as space.

There is nothing beyond the universe as the word 'universe' means everything that is. As the universe expands it creates space. There is no space beyond the universe. This idea is very hard to imagine as it is at the very edge of what humans can comprehend.

Sizes and distances in the universe are also difficult to visualise. The distances between objects in the universe are so huge that we measure them in **light years**.

The expanding universe

The universe continued to grow explosively. Its temperature was very hot, reaching millions of millions of degrees. It was far too hot for matter as we know it today to exist. As the universe cooled down, sub-atomic particles 'condensed' out of the cooling energy and atoms followed. All the atoms organised themselves in new ways, according to rules and regulations that they still obey, and that we are still discovering.

Ideas about the end of the universe

- The big question is **whether the expansion of the universe will continue?** The only thing that can stop it is gravity. If there is enough mass in the universe, gravitational attraction will eventually slow the expansion and cause contraction. The universe will shrink and the galaxies will come closer together. As they do the strength of the gravitational attraction will increase and this will speed up the contraction. So in a few hundred billion years, long after the sun is burnt out, the universe may end in a **'Big Collapse'**! It could be that the Big Collapse is followed by another Big Bang and a new universe is created — there is of course no scientific evidence for this.
- There may actually be no Big Collapse. If there is not enough matter in the universe for gravity to overcome the initial incredible explosion, **movement may continue outwards**. Clusters of galaxies may continue to fly apart and the universe will continue to expand. Even though most people think that this is what will happen, physicists believe that it will eventually come to an end. The stars will stop shining and concentrations of energy and matter will dissipate further and further apart until only a frozen darkness remains.

- Another possibility is for gravitation to perfectly balance expansion. Then the outward movement becomes slower and slower but never stops completely.

And in the end ...

At the moment most scientists believe that the universe will expand forever. Exciting research carried out by astronomers suggests that the expansion of the universe is faster than they thought and is in fact speeding up. Einstein suggested that there is a force pushing matter apart and therefore opposing gravity. This idea would guarantee that expansion continues. Scientists need more and more information before they can find out for sure what will happen.

Cosmological contents

As well as trillions of stars and lots of gas the universe contains some very curious objects.

Black holes

A black hole is an object with such great mass (and therefore enormous gravitational pull), that nothing can leave it, not even light. Black holes can form when the core of a massive star collapses in on itself. In the centre of the black hole the original material is compressed to an incredibly high density. The force of gravity is great enough to swallow light and space.

Neutron stars

These are incredibly dense objects. Neutron stars form after a supernova occurs (a giant star exploding). After the explosion, the core of the star collapses under its own gravity, forcing the electrons into the nucleus of the atoms, forming particles called neutrons. A teaspoon of neutrons would weigh a billion tonnes!

Pulsars

These are special types of neutron stars that emit beams of radio waves, and also spin. As the pulsar spins, its beam sweeps in our direction and we detect a pulse of radio waves. These signals are very regular. The CSIRO's Parkes telescope holds the record for finding pulsars.

Quasars

Quasars are the bright cores of distant galaxies. The Anglo-Australian telescope at Coonabarabran in New South Wales has observed more than 10,000 quasars, which is helping astronomers to work out the structure of the universe.

Dark matter

Astronomers know that there is more matter out there than they can recognise. Gravity comes from matter but the amount of matter scientists can detect is not enough to account for the amount of gravity. Therefore, there must be a lot of matter out there that does not radiate light. This is called 'dark matter' and scientists believe it makes up most of the universe.

Gravitational lenses

If a light from a distant source passes close to a very massive object, the rays passing near the object will bend. This is called gravitational lensing. Astronomers can use gravitational lensing to study distant objects that they could not normally see as lensing makes their images brighter.

Where did our solar system come from?

Most astronomers believe that all the planets were formed at the same time as the Sun — about 4 500 million years ago.

Scientists say that the solar system began with a huge cloud of cosmic dust, ice, hydrogen and helium gas called a **nebula**. Due to its own gravity, the cloud began to shrink. As it got smaller, particles fell inwards and the centre became hotter and hotter. Eventually the centre of the cloud was white hot and **nuclear fusion** began. It became the **star** we call the **Sun** and the planets formed from concentric rings around the Sun.

Nuclear Fusion occurs when hydrogen nuclei fuse together and release energy. An example of this type of reaction follows.

Hydrogen → (Nuclear Fusion) → Helium + Energy

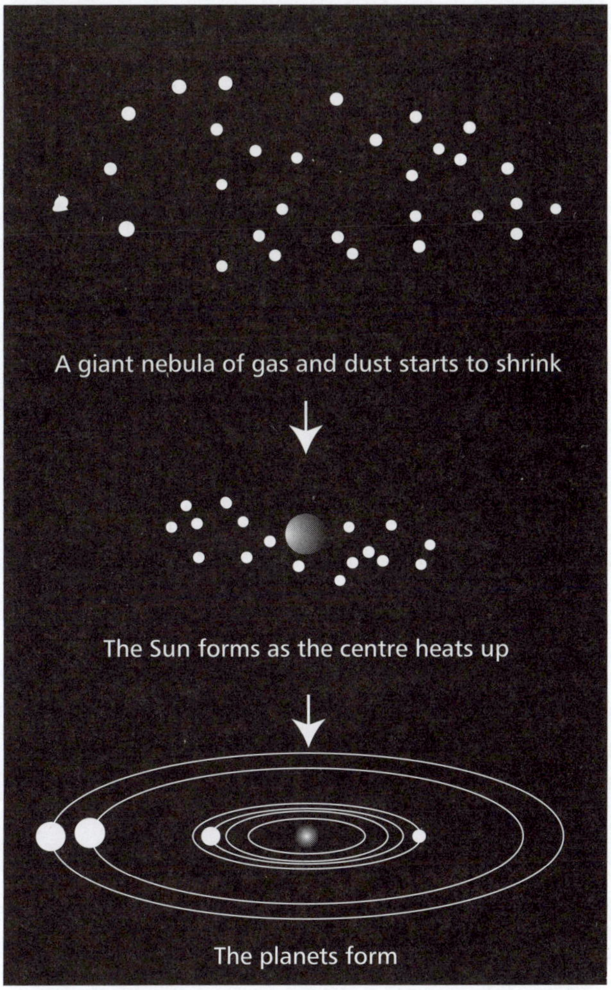

The formation of the planets

The dust began to stick together forming four rocky inner planets. The outer planets formed from the ice and gas to become the 'gas giants' like Jupiter.

Asteroids are some of the rocky bits left over and they orbit between Mars and Jupiter. **Comets** may also be leftovers. They are lumps of ice with wispy tails of gas and dust. They have elliptical orbits that bring them close to the Sun and then far out into the solar system.

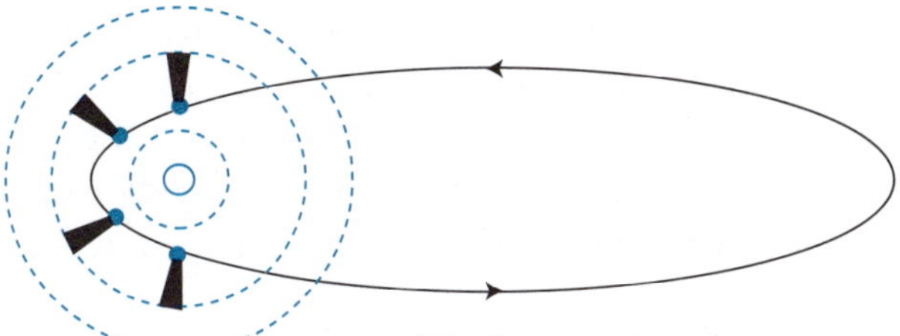

Orbit of a comet

Our star the Sun

Stars are a very long way away. We can see them because they send out huge amounts of light energy. Our Sun is a medium to small sized star. Sirius is the next brightest star in our sky. It is much bigger than the Sun, sending out much more light but it is a long way away. Groups of stars that form patterns are called **constellations**, for example, the constellation of Orion and the Southern Cross.

The Sun consists of several layers surrounding a core. It is made mostly of hydrogen and helium. Overall the Sun is not dense but the matter in the core itself is very densely packed.

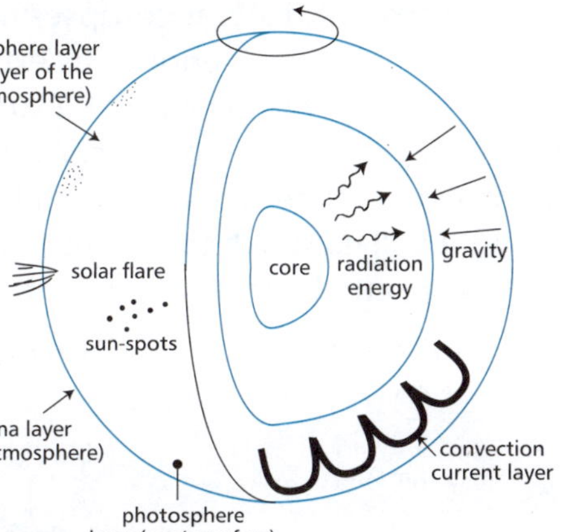

Structure of the Sun

The Sun is using up its mass at the rate of 4 million tonnes each second. It has been doing this for 4 500 million years, but is still only halfway through its lifetime. The huge amount of energy radiated out eventually arrives on the earth and keeps us alive.

Inside the Sun there is a constant battle of forces — the pull of gravity is trying to crush the sun, while the energy from the very hot core is trying to make it expand. As long as these forces remain balanced the star is stable.

Sunspots are slightly cooler areas on the Sun's surface. These sunspots come and go as the power of the Sun varies. **Solar flares** are the result of intense magnetic activity. They can spew gas more than 300 000 kilometres into space.

The life and death of a star

Stars eventually use up their fuel and start to change. This change depends on how big the star is.

- Small stars like our Sun are called **yellow dwarfs**. After about 10 000 million years the hydrogen is all used up and the star starts to use the helium as fuel. This makes heavier elements like oxygen and carbon. As it does this it expands and cools to become a **red giant**. When it has used up all its fuel it collapses under gravity to form a very hot and dense **white dwarf**. Finally it cools to become a **black dwarf**.
- A larger star burns hotter (a **blue giant**) and runs out of fuel faster. It expands and cools to a **red supergiant**. Later it collapses and blows off its outer layer in an enormous explosion called a **supernova**. The core of a supernova finally collapses to become a very dense **neutron star**. If the star was much bigger that the Sun it will continue to collapse under its own gravity and become a **black hole**.

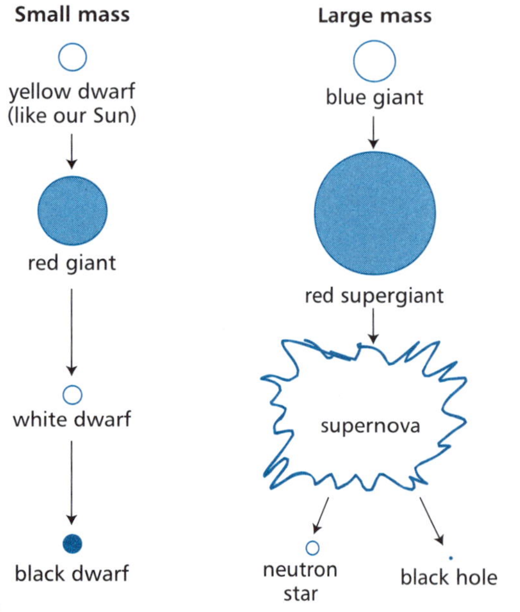

Death of dwarf and giant stars

The Milky Way

Our galaxy is a collection of about 10 000 million stars; our Sun is just one of them, positioned somewhere near the edge in a spiral arm. The Milky Way rotates around its own centre.

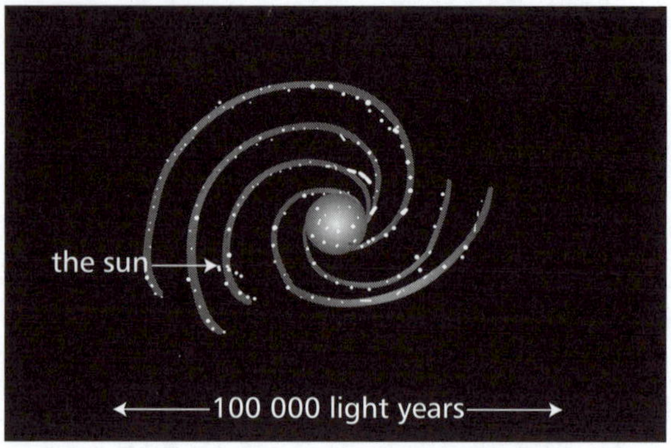

The Milky Way

The Galaxy is huge. It takes eight minutes for light from the Sun to travel from the Sun to the Earth, four years for light to travel from the nearest star, and 100 000 years for light to travel right across our Galaxy.

Are there other solar systems like ours? Could there be Earth-like planets and life elsewhere in the universe? Recent research tends to support the idea that life could have developed somewhere else in the universe.

Glossary

Asteroid
A small, rocky object revolving around the Sun.

Big Bang
A popular name for one of many theories explaining the origin of the universe.

Black hole
A region containing a huge amount of compacted mass, making its pull of gravity so strong that nothing, not even light, can escape from it.

Comet
A ball of rock and ice, usually a few kilometres across, with a long wispy tail of gas and dust which follows an elongated orbit around the Sun.

Cosmology
The study of the universe on the grandest scale, especially its structure, origin, evolution and density.

Dwarf star
Any star comparable to or smaller than the Sun.

Galaxy
A collection of billions of stars and matter held loosely together by gravity. Galaxies are often spread over thousands of light years. The Milky Way is an example.

Giant star
Any star much larger than the Sun.

Light year
The distance travelled by light in one year. Travelling at 300 000 km per second, light travels almost 9 461 trillion km in one year.

Mass
A measure of the total amount of matter contained within an object.

Nebula
A contracting, swirling cloud of dust, ice and gas that forms a concentric series of rings from which the planets were formed.

Nuclear fusion
The reaction whereby hydrogen atoms join together to form helium. Each time this happens a large amount of energy is released.

Star
Distant suns that give out their own light.

Sun
The star at the centre of the solar system.

Solar flare
Violent magnetic activity that releases pent-up energy.

Sunspots
Sunspots appear as dark spots on the surface of the Sun.

Question 1

Look at the drawings below. If they are placed in the correct order, each one will be 1000 times wider than the one before it. Unscramble the drawings and arrange them from the smallest to the largest.

A

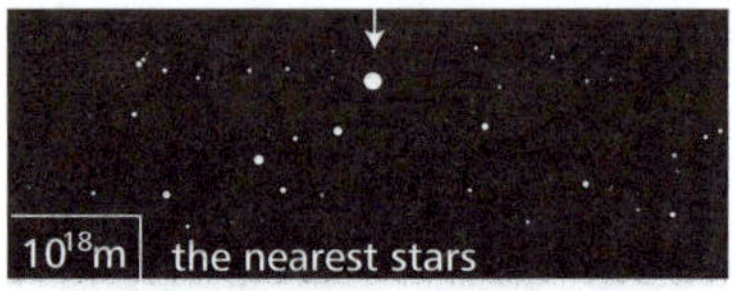

B

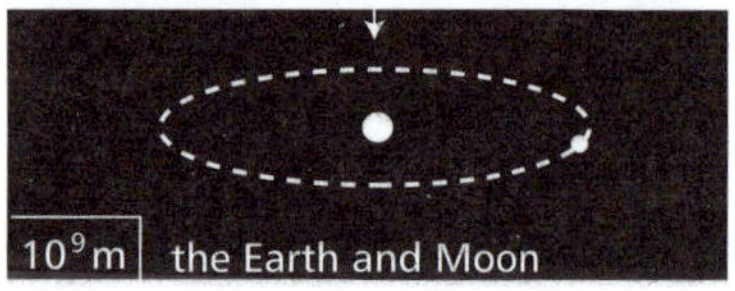

C

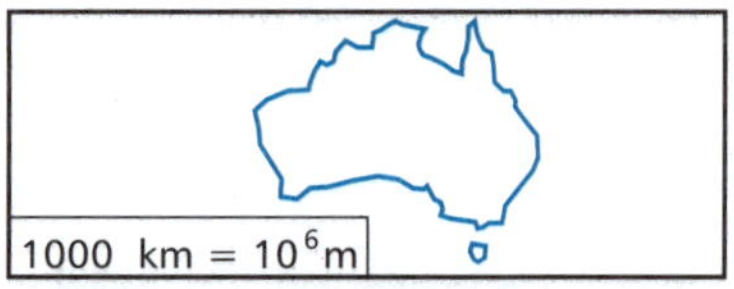

D

E

F

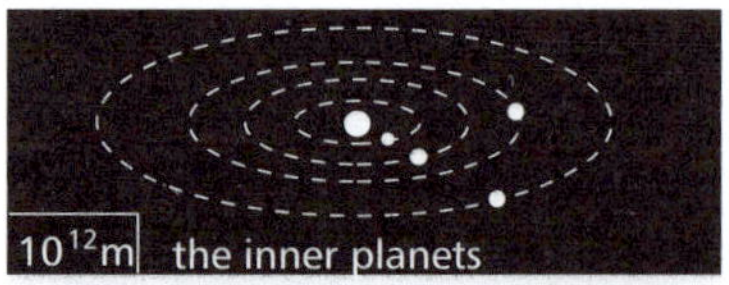

G

H

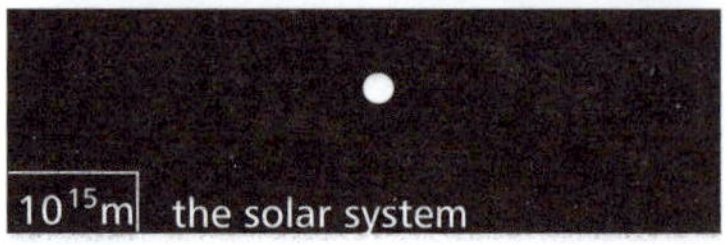

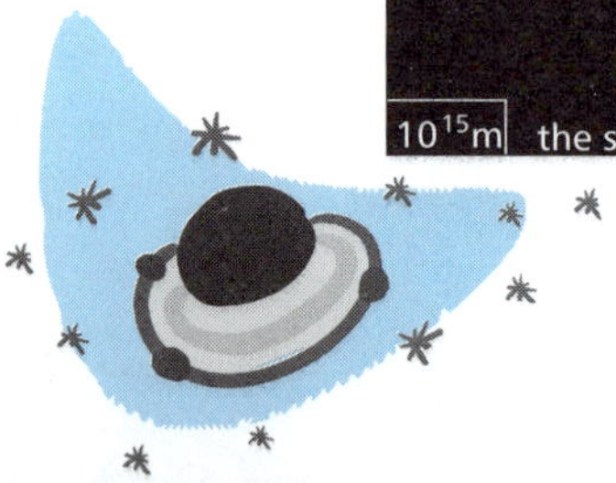

Question 2

Label the diagram of the Sun below with the following structures:

core, solar flare, sunspots, radiation energy, gravity.

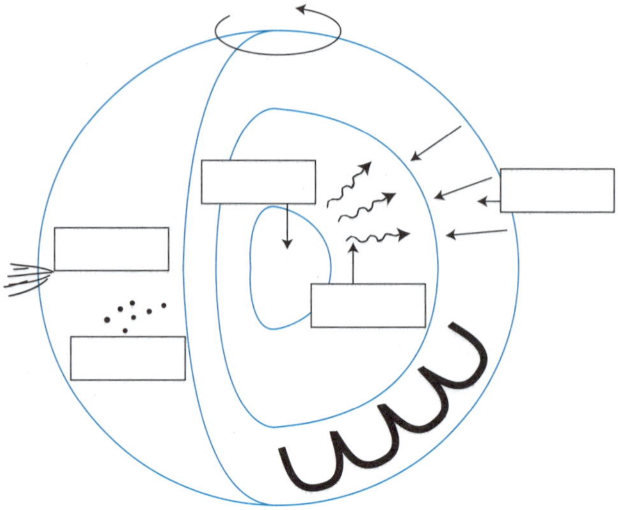

Question 3

The following diagram displays the life and death of three different types sized stars. Use the descriptions below to correctly label the three deaths.

a Collapses to a black hole.

b Contracts to a white dwarf.

c Becomes a supernova then collapses to a neutron star.

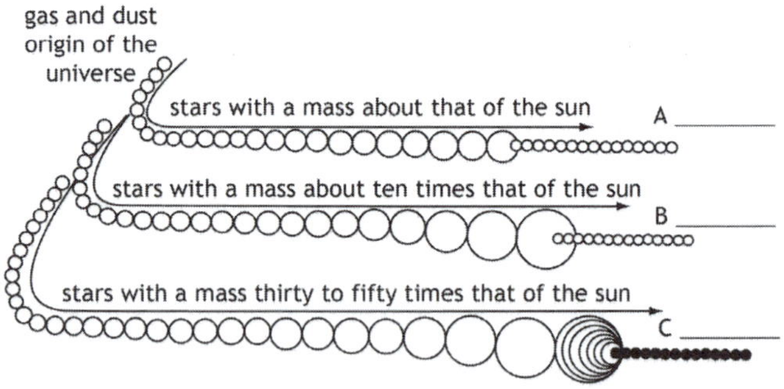

Question 4

The diagram below is a section through to the centre of the Sun. Temperature is measured in Celsius (°C). Draw a histogram to show the temperature of the various layers of the Sun.

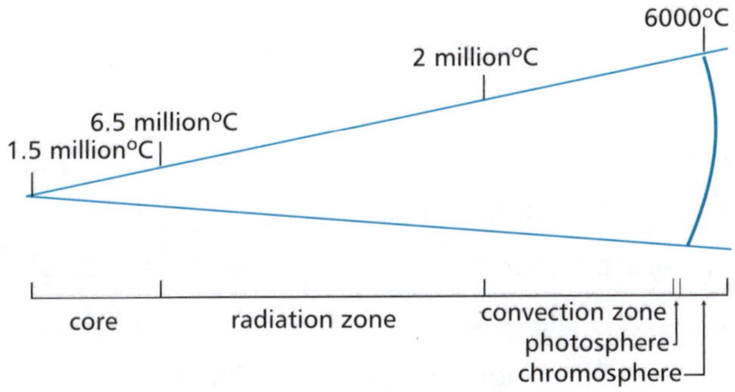

Question 5

Sunspots occur in cycles of about 11 years. As shown in the graph below the average number of sunspots varies; periods of unusually large numbers are called sunspot maximum, periods with unusually small numbers, sunspot minimum.

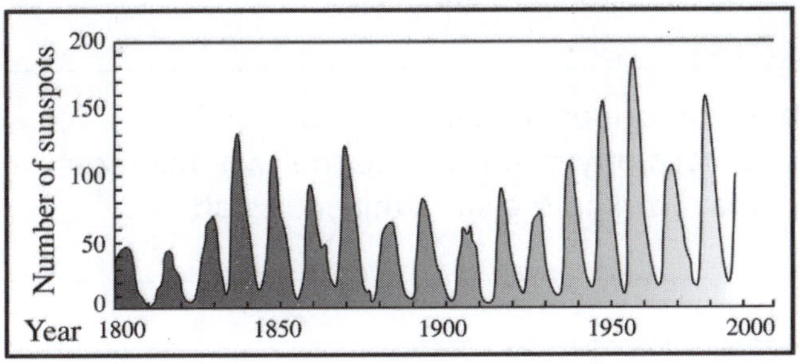

a How many cycles were sunspot maximums (150 or above)?

---

b How many cycles were sunspot minimums (50 or less)?

Question 6

a Write in the correct terms to describe the stages in the death of a small star like our Sun.

b Write in the correct terms to describe the stages in the death of a large star.

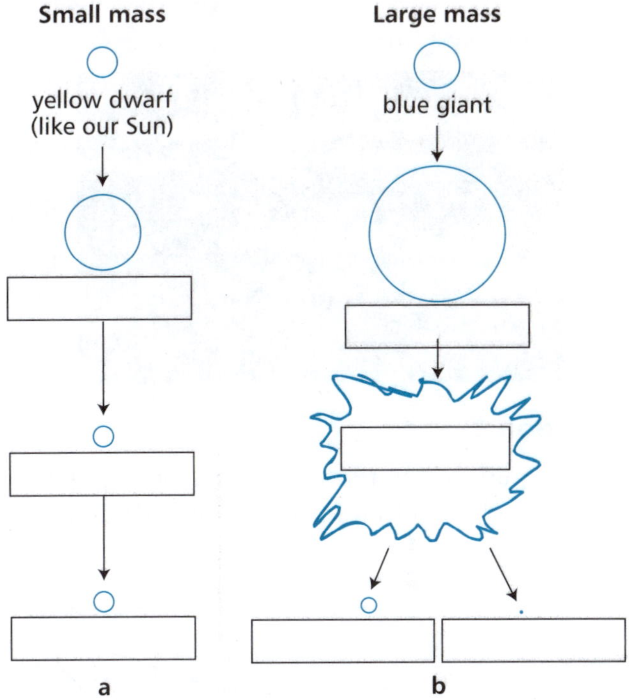

Question 7

Different animals live in different environments to stay alive. Our telescopes are not strong enough for us to be able to see plants and animals on other planets so scientists make guesses based on the information that they do have. Imagine a planet that is just on the edge of what scientists call the 'livable zone' of its sun. This is where scientists think life is most likely to occur. The planet has the following characteristics:

Planet name	Freezeland
Precipitation (rainfall)	High
Average temperature	–10°C
Moons	2

Which of the following forms do you think life on this planet would take? Explain your prediction.

a A long thin shape, like a worm.

b A spherical shape, like an orange.

c An irregular shape, like a star.

d A long shape with lumps at each end, like a dumb bell.

Question 8

The diagram below is most likely to be:

a A black hole.

b An exploding star.

c A galaxy.

d The solar system.

Question 9

The diagram below represents the solar system. Which labelled path represents the path of a comet?

____

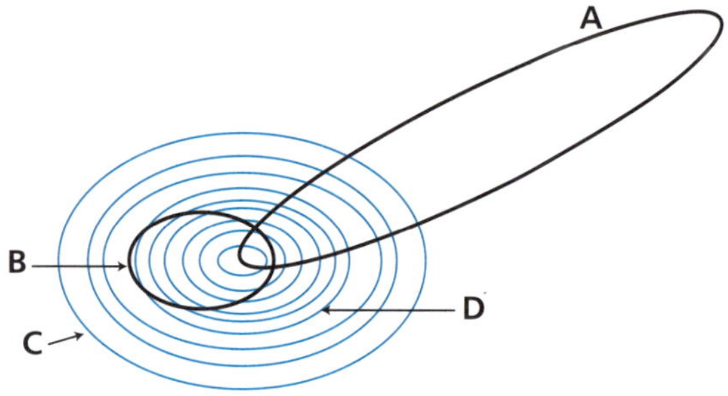

Question 10

Look at the scale diagram below. The Big Bang is at the bottom and the universe is continually expanding towards the top until it eventually stops. We are at present at point 'X' on the way out. How far has the universe expanded?

a Halfway.

b Almost all the way.

c One third of the way.

d One quarter of the way.

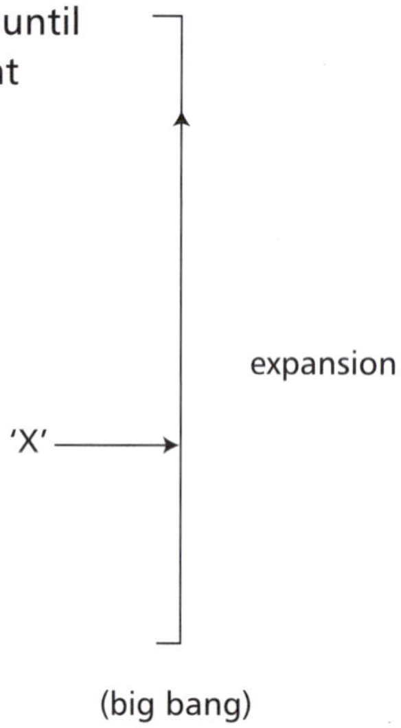

Revision questions

Question 1

Fill in the missing words:

All the stars except the ________ are a very long way away. Like the Sun, they give out their own _______. The brightest star in the sky is _______. This star is much ________ than the Sun and gives out _______ light but it is much further away.

Question 2

Stars are formed from massive clouds of dust and gases in space.

a What force pulls the clouds of dust and gas together to form stars?

____

b Once formed, a star can have a stable life for billions of years. Describe the two main forces that work in the star while it is stable?

____

____

c What happens to a star once this stable period is over?

____

d What shape are the orbits of comets?

____

Question 3

Our Sun is a star. It looks much brighter than other stars because:

a It is brighter.

b It is closer than other stars.

c It is yellow when other stars are white.

d The other stars are in other galaxies.

Question 4

A group of stars forming the Southern Cross is an example of a:

a galaxy

b constellation

c solar system

d nebula

Question 5

The energy produced by stars is most likely the result of:

a combustion of hydrogen gas

b gravitational energy

c nuclear energy

d potential energy

Question 6

Which of the following theories for the end of the universe do most scientists support at the moment?

a The Big Bang followed by the Big Collapse — the universe will expand but finally collapse back on itself.

b The Big Bang — the universe will continue to expand until there is just a frozen void left.

c The expansion will be balanced by gravity.

Question 7

a Fill in the missing words:

Our Sun is a _______ dwarf star. Later it will change to a red ________, followed by a ________ dwarf, then a ________ dwarf.

b A larger star (a blue _______) will change to a red _______ and later explode in a ______ leaving a ______ star, or a _______ hole.

Question 8

Provide the cosmological objects with the correct explanation:

Object	Explanation
Black Hole	
Gravitational Lens	
Quasar	
Pulsar	
Neutron Star	
Dark Matter	

Question 9

Refer to the diagram showing the structure of the Sun on page 61, and list the four outer layers of the Sun.

____

____

____

Question 10

Asteroids are made of:

a Ice and gas left over from the solar system.

b Energy from exploding stars.

c Dust.

d Rocky bits left over from the formation of the inner planets.

Question 11

Distances in astronomy are extremely large so they are measured in light years. This is the distance that light can travel in one year (9 461 trillion km).

a The nearest galaxy to us is called the Large Megellanic Cloud. It is only 160 000 light years away. Could a space traveller reach this galaxy within his lifetime?

____

b This galaxy is 30 000 light-years in diameter. How many trillion km wide is it?

____

Question 12

Complete the sentences below:

The Sun is a star in our galaxy, the _______ _____. The Sun's energy comes from _______ ______. A star is formed from a _________, and changes as it gets older. The universe continues to _________.

Question 13

a Explain the process of nuclear fusion.

____

____

____

b Write the equation for one example of nuclear fusion.

____

Question 14

a Where in the solar system is the asteroid belt?

____

b Do asteroids move around the Sun in orbits?

____

c What is the difference between a comet and an asteroid?

____

____

Question 15

Scientists think that the Sun formed about five billion years ago. Much later the Earth formed and then life began. Look at the time line below:

Sun formed

Earth formed now

Life began

*___*_____*_______________________

5 4 3 2 1 0

Billions of years

a When did life on Earth begin?

____

b About how long ago did the Earth form?

____

c How many more years will the Sun keep on shining if it has so far used up half its mass?

____

CHAPTER 7

HUMAN IMPACT ON THE BIOSPHERE

Which species is headed for extinction next?

The major problems facing the environment today have been largely caused by humans and their activities. Humans use up a large share of the Earth's resources and create vast quantities of harmful waste and pollution.

Humans in the system

- Humans are a recent part of life on Earth.
- Human intelligence is a unique characteristic that is responsible for the scale of human impact on the environment.
- Humans are not specialised for any particular niche but they can make tools, use fire and they work together. They can cope with almost any terrestrial habitat and have increased their range to cover most of the planet.
- The first Australians were hunters and gatherers. They tended to live with the environment, moving around and allowing each patch of land to recover.
- Agriculture interferes with the ecosystem and diverts energy and nutrients to make biological products chosen by humans.
- Agriculture allowed urbanisation to occur. Cities have expanded and control the land around them, destroying the natural environment. People have moved from agricultural to business and industrial occupations.
- The Industrial Revolution produced the technological power to rapidly make large changes to the environment.

The industrial revolution

The Industrial revolution in the 1800s made a major contribution to the world we know today. Industrial processes based on fossil fuels (non-renewable energy sources), for the first time permitted humans to significantly alter the environment.

The main features were:

- The application of **new technology** to both urban and rural industries, for example: steam engines, electric power, internal combustion engine, transport and communication.
- The replacement of human and animal labour by **machines**. Machines are powerful and fast, and therefore more efficient.
- The replacement of **renewable** sources of energy and materials by **non-renewable** ones, for example, fossil fuels.
- A vast increase in **technological knowledge**, for example, scientific knowledge, hygiene, clean water and safe disposal of waste.

The main consequences of the Industrial Revolution have been:

- A great increase in the use of energy and resources.
- A great increase in the production of waste such as air **pollution**, toxic and heavy metals and the quantity of household waste.
- A significant increase in the standard of living for a large part of the population. Modern gadgets save time and allow for increased leisure time.

- Rapid and continuous economic growth. Nations build more, mine more and produce more, so there is more wealth.
- A sudden growth in population. Knowledge of hygiene and medicine, together with increased efficiencies in farming and manufacturing, allow more people to live longer.
- Growth of cities. Cities have grown to enormous sizes, consuming vast resources and producing huge amounts of waste.
- An increase in the destructive power available for warfare.
- Widespread changes to the natural environment. The chemicals released by industrial activity have polluted the sea, the air and the land. **Biodiversity** is decreasing and population is increasing.

All societies in the past have had an effect on the natural environment. The difference today is in the numbers of people and the power available to them, hence the scale and speed of the damage.

The global effect of modern humans

With our huge population and our use of science and technology, our impact on the planet is increasing.

The atmosphere

The atmosphere makes Earth unique — it provides our weather and protects the **biosphere** from sun damage. Some of the most extensive, but least visible, human impact has been on the atmosphere.

- **Acid rain** — rainwater dissolves gases in the atmosphere. If the atmosphere contains pollutant gases, such as sulphur oxides from power stations, carbon dioxide from cars, and nitrogen oxides from factories, the rain becomes highly acid. This rain can kill forests and dissolve some rocks.

Acid rain

- **Ozone depletion** — the ozone layer is a band of gas (a form of oxygen) in the atmosphere, about 40 kilometres above the ground. It protects the Earth from some of the harmful ultra violet radiation in sunlight. Recently, synthetic gases such as chlorofluorocarbons (CFCs) have built up in the atmosphere. These gases react with the ozone, making parts of the ozone thinner.

  Ozone depletion allows more UV radiation to reach the earth. This is dangerous for many living things as UV can damage genes, causing cell malfunction, which can lead to cancer.
- **The greenhouse effect** — when the sun's energy reaches the earth, some energy is reflected back and some is trapped in the atmosphere by gases such as carbon dioxide. These gases keep the atmosphere warm enough to support life. Due to deforestation and the burning of fossil fuels there has been an increase in the amount of carbon dioxide in the earth's atmosphere. Many scientists believe this has caused an increase in the earth's average temperature. If **global warming** continues it will have devastating effects on the earth's ecosystems.

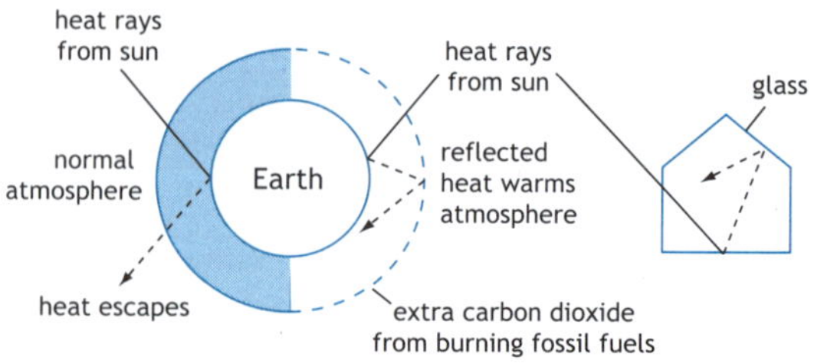

Greenhouse effect

The land

Agriculture is the main agent of change to the Earth's surface.

- **Deforestation** — many parts of the world have been completely cleared of trees to make space for farming. This deforestation often results in **desertification**, a process that results in the mineral nutrients in the soil being washed away, turning fertile land into semi-desert. Loss of forest causes decreases in biodiversity, changes to the world's climate and increased levels of greenhouse gases.
- **Soil erosion** — this is the process of soil being washed or blown away by water and wind. It is the most common and the most devastating form of land degradation. Clearing the land, ploughing and overgrazing destroys the soil structure and the plant roots that hold the soil in place — the soil and its nutrients are lost.
- **Salinity** — clearing deep-rooted trees can result in the ground forming a watertable, and its dissolved salts rising to the surface. Salt accumulation occurs in root zones and on the surface in low-lying areas. Irrigation makes the situation worse. Salting of the land causes severe loss of productive land.
- **Waste** — biodegradable substances in waste materials, such as paper, will be broken down by decomposers into harmless substances and used in the environment. Many plastics, however, are non-biodegradable, cannot be broken down and go on to accumulate and pollute the land for many years. Also, wild animals and sea creatures may try to eat it, or become tangled in waste plastic, and often perish.
- **Biocides** — biocides are chemicals that kill living things, for example herbicides kill plant pests. These chemicals are good for farming but they also destroy other living things. Biological control, however, makes use of natural predators that attack the pest without harming other species or leaving dangerous chemicals in the environment.

Water

In many parts of the world water is a scarce resource.

- **Silting** — once trees have been removed from the ground, water flows very quickly over that land, carrying the soil with it causing flooding and the silting up of creek and river systems.
- **Salinity** — salt on the land is washed into rivers and creeks. The increased level of salt in the water kills many aquatic species and reduces aquatic biodiversity.
- **Eutrophication** — is the process by which all the oxygen in water is used up. Many of the waste substances produced by human activities contain mineral elements, such as nitrates and phosphates. If these are washed into water they act as a fertiliser, and may cause algal blooms. The overgrowth of algae uses up the oxygen in the water, killing other forms of life through oxygen deprivation.

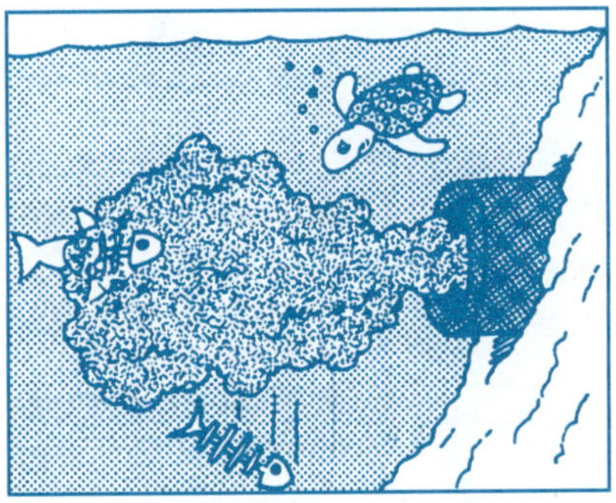

Eutrophication

Oceans

Oceans, like the atmosphere, are used by many countries, but are the responsibility of no one.

- **Marine pollution** — the main pollutants are excess nutrients, synthetic chemicals, plastic, oil, heavy metals and radioactive waste. These pollutants enter the sea via water run-off, sewage, oil spillage from tankers and garbage dumping.

- **Habitat destruction** — coastal habitats such as mangroves and sea grass, which are used by vast numbers of marine creatures for reproduction and as nurseries for their young, are destroyed by human habitation.

Ecological sustainability

Sustainable development is development that is compatible with the rest of the environment, such that the basic needs of all are met.

The key principles follow:

- Intergenerational equity — the present generation should leave the next generation a healthy, diverse and productive environment.
- Conservation of biodiversity and ecological integrity. This should be a constraint on all economic activity.
- Constant natural capital and sustainable income. Only income that can be sustained indefinitely should be taken.
- Precautionary policy approach. Decisions should be cautious until proof of sustainable development is received.
- Pricing environmental values and natural resources to recover the full social and environmental cost.

Conservation

Many human activities damage other forms of life. Conservation is the careful maintenance, management and preservation of living things and natural resources to make sure that we harm the natural world as little as possible. Examples:

- Retain natural environments such as parks and forests.
- Use renewable energy resources.
- Recycle wherever possible.
- Reduce pollutants.
- Reduce our dependency on chemicals.
- Control human population growth.

Glossary

Acid rain
Rain that has been made strongly acidic by atmospheric pollutants.

Biodiversity
The variety of species in an area.

Biosphere
All parts of the earth that make up the living world.

Conservation
The management of resources to protect the natural world.

Deforestation
The permanent removal of trees.

Ecology
The study of the interactions of organisms with their physical environment and with one another.

Eutrophication
The increase in the nutrient content of a body of water, leading to a rapid increase in algae growth. When this algae dies the oxygen level in the water drops until it cannot support other aquatic life forms.

Global warming
An increase in the earth's average temperature due to deforestation and burning fossil fuels.

Greenhouse
The trapping of heat by the earth's atmosphere — like the glass in a greenhouse.

Pollution
The disruption of the natural world by the release of chemicals or other agents.

Salinity
The concentration of salts in the root zone by evaporation from a shallow water table.

Science skills

Question 1

The following diagram shows the process of deforestation, farming and land degradation.

List the diagrams A to D in the correct order to show the process.

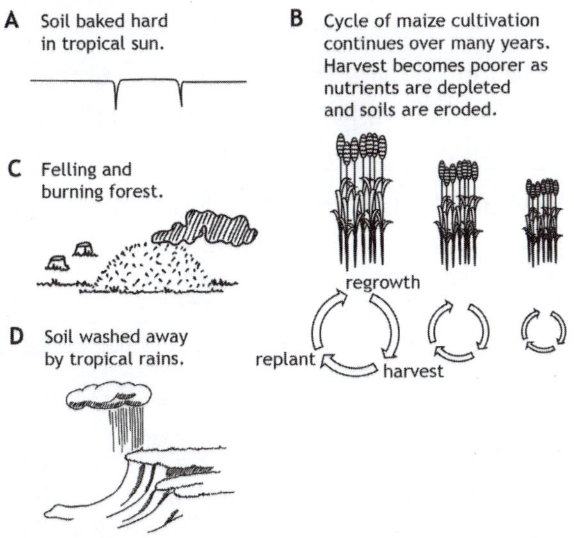

Question 2

Some students carried out an experiment on acid rain. They put a little tap water (pH = neutral) into a jar and burnt some matches in the jar and then screwed on the lid.

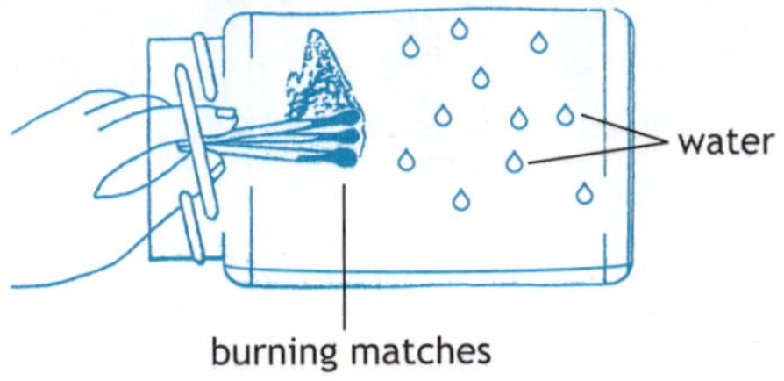

a They tested the water for acidity. Would the pH be acid or alkaline?

____

b Explain your answer.

____

____

Question 3

The following data refer to a population of micro-organisms in a culture.

Time (days)	0	2	4	6	8
Microbes per mL	30	160	600	620	400

a Graph this data.

b What is happening to the micro-organisms between days 2 and 4?

____

c What is happening between days 4 and 6?

____

d What is happening on day 8?

____

____

Question 4

Air pollution is a major source of environmental damage. Complete the table below, which summarises the source and effects of pollutants.

Pollutant	Source	Release	Effects
	domestic industrial	aerosol sprays	destroys the ozone layer
sulphur oxides	domestic industrial		acid rain
nitrous oxides	domestic industrial	burning fossil fuels	
carbon dioxide	domestic industrial	burning fossil fuels	
heavy metals		car fumes	accumulation in plants and animals

Question 5

Study the graph below, which shows the demand for natural resources.

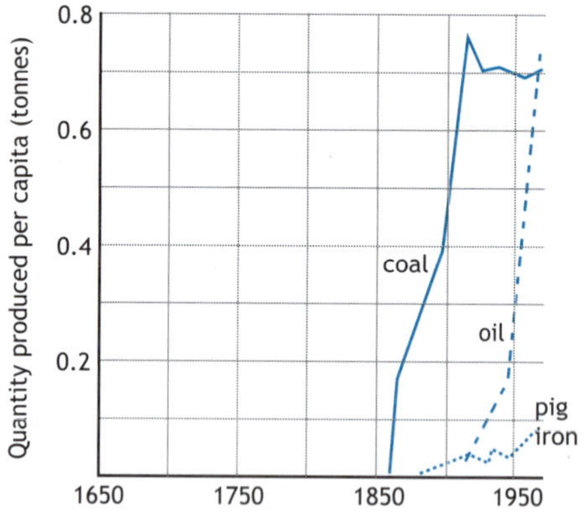

a After what year was there an enormous increase in the demand for fossil fuels?

____

b What revolution did this coincide with?

____

Question 6

Students studied the effects of salination on the germination of bean and sunflower seeds. They set up the following experiment:

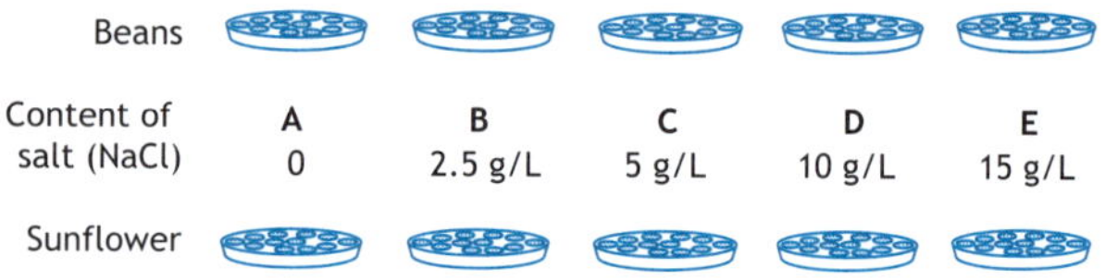

Germination rates were as follows:

Salt level	Beans	Sunflowers
A = 0	100%	100%
B = 2.5g/L	80%	90%
C = 5 g/L	50%	80%
D = 10 g/L	30%	70%
E = 15 g/L	0	50%

a Which seed is most susceptible to salt?

____

b Which crop would be best planted in slightly salty soil?

____

Question 7

The graph below shows the increase in world population from 10 000 years ago until the present.

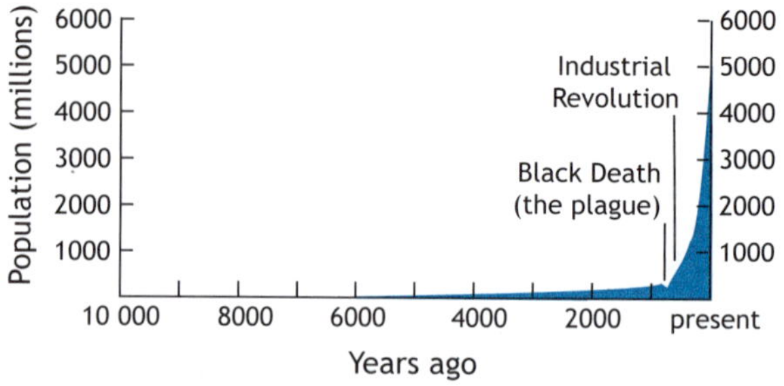

a The most spectacular growth has been over the past how many years?

____

b How long do you think it will be until there are another 1000 million people on the planet?

____

c Why did the Industrial Revolution trigger an increase in human population?

____

____

____

Question 8

The graph shows the energy released and the carbon emitted in Australia for the four fossil fuels listed.

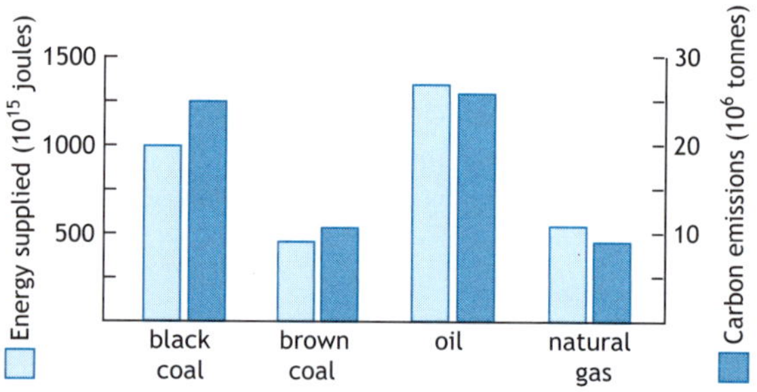

a Which two fuels produce the greatest carbon emissions?

____

b Which fuel produces the least energy?

____

c Rank the fuels according to their greenhouse potential.

____

d Explain why some fuels account for more of our energy needs than others?

____

____

____

Question 9

The following table outlines the flow of resources into and out of three different ecosystems. Use the words *little*, *some* or *much* to describe the resource flow.

Resource	Urban	Agricultural	Natural
energy in			
energy wastage			
materials in			
recycling			

Question 10

The following diagram shows water movement in a healthy environment.

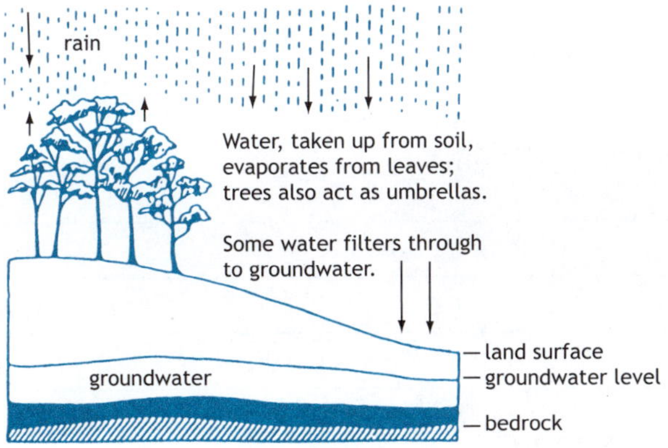

In the diagram below fill in the boxes with the correct statement from the list below to explain the development of salinity in soils.

- trees removed
- salty water emerges
- ground water rises, dissolves salt
- more water filters through to ground water

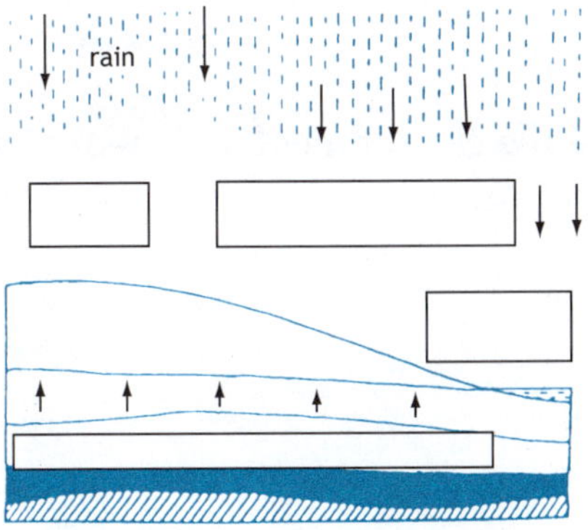

Question 11

Study the diagram below of the greenhouse effect and write down the names of the gases that contribute to global warming.

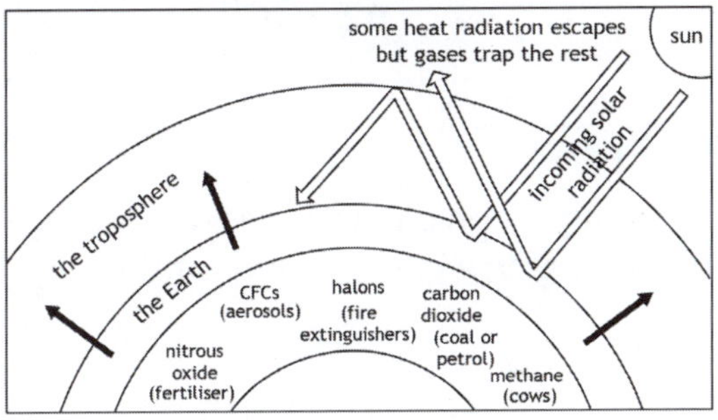

Activities that release gases and contribute to greenhouse effect

Question 12

The two graphs below show carbon dioxide and methane concentration in the air over the past 200 years.

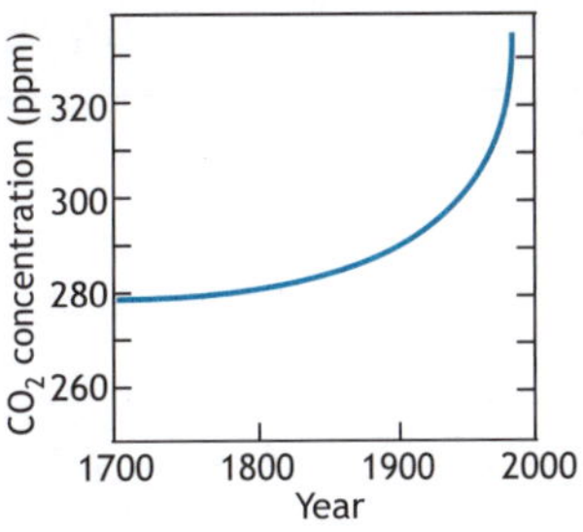

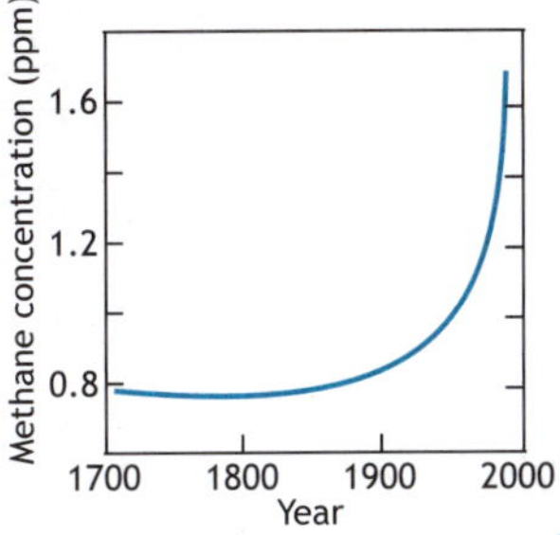

a Why has their been such a large increase in carbon dioxide in recent years?

b Why has there been such an increase in methane production?

Question 13

What are the environmental consequences of the following situation?

Question 14

The following graph shows levels of ozone and nitrogen oxides in a city throughout the day.

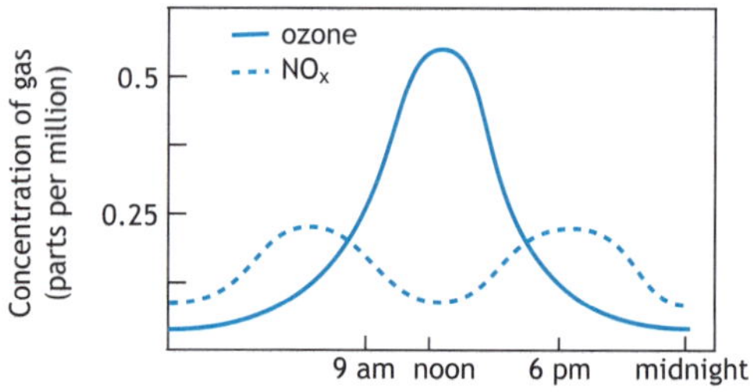

What is the source of nitrogen oxides?

Why are their two peaks?

Why does ozone peak in the middle of the day?

Revision questions

Question 1

What are the special features of *homo sapiens* that make it such an adaptable species?

Question 2

Look at the diagram of the chief gases in the enhanced greenhouse effect. Decide which gas accounts for just over half of global warming.

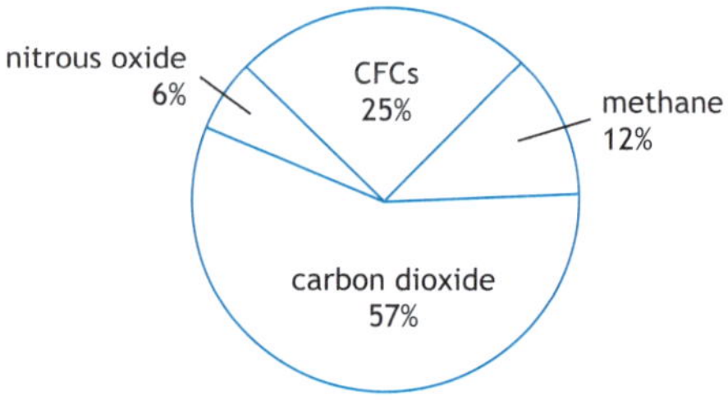

Question 3

List some of the predicted consequences of global warming.

Question 4

How has recent human activity shortened the storage time of the carbon stored underground?

Question 5

How is deforestation on hillsides related to erosion and flooding?

Question 6

Why are more and more people living in cities?

Question 7

List in the table below some of the inputs and outputs of cities.

Inputs to a city	Outputs of a city

Question 8

Complete the table below by listing the effects of ocean pollutants.

Pollutant	Effect
heavy metals	
crude oil	
plastics	
sewage	

Question 9

The main effects of eutrophication are as follows: species diversity decreases, plant and algal biomass increase, turbidity increases and sedimentation increases.

What are some of the problems this will cause humans?

Question 10

The four stages of human development are listed below. Put them in the correct order and match the comment about man's position in the environment with each one.

Agriculturalist	Great control over the environment but with many undesirable consequences.
Hunter gatherer	Controlled by the environment.
Industrialist	Man has increased control over the environment.
Primitive man	Man starts to control the environment.

Question 11

Overgrazing and trampling of grassland around oasis water source, eg. North Africa.

a What are the environmental consequences of the following situation?

b What are the human consequences?

c How could the management of this grazing land be improved?

Question 12

The Industrial Revolution made a major contribution to the world we know now. List 6 of the major consequences.

Question 13

Fill in the missing words.

Continuous growing of crops eventually uses up all the soil's ________. These can be replaced by artificial ________ but the cost to the ________ is high. Fertilisers can ________ rivers and add to ________ problems.

Question 14

The invention of the car was one of the biggest changes that occurred in the last 100 years.

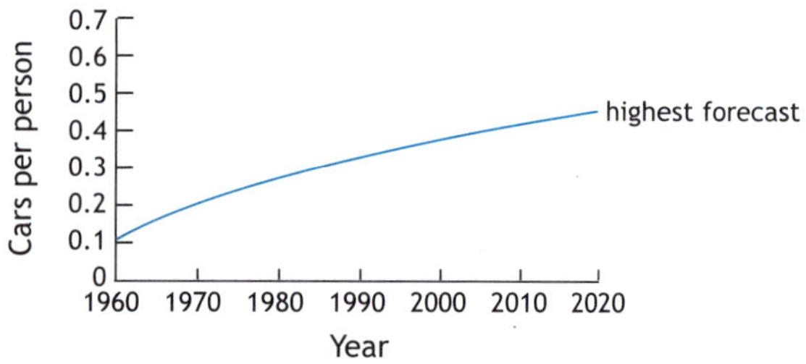

a What does the graph tell us about the use of the motor car?

b Do you think pollution from cars will get better or worse in the future? Why?

c By the year 2020 how many cars will every person in the world own?

Question 15

Fossil fuels and resources are:

a non-renewable

b renewable

c recycled

d exponential

Question 16

The energy released when some fuels are burnt is shown in the table below. Which fuel releases the most energy per gram?

Fuel	Amount of fuel burnt (g)	Energy released (kJ)
hydrogen	2	285
carbon	12	393
methane	16	890
ethanol	46	1 360
octane	114	5 460

Question 17

a What damage does ultra violet radiation do to living things?

b What synthetic gas causes a thinning in the ozone layer?

Question 18

All the waste we produce has to go somewhere here on earth.

a What happens to waste that seems to disappear when you burn it?

b Where does the waste that we flush down the toilet finish up, treated or not?

c How can we deal with our waste so that it does not damage the environment?

Question 19

a What is sustainable economic growth?

b List two key principles of sustainable economic growth.

Question 20

a What is conservation?

b List two examples of conservation.

CHAPTER 8

WAVE MOTION

Why do ocean waves break?

Energy is often moved from one place to another by waves. Waves can be either **mechanical** or **electromagnetic**. There are two types of waves — **transverse** and **longitudinal**.

Mechanical waves

Mechanical waves disturb material. The particles in the material oscillate or vibrate and transfer energy. The particles could be in a solid, a liquid or a gas.

Example: a wave of energy passing along a rope is a mechanical wave. The rope itself does not move along, only energy moves in the direction of the wave's travel. After the wave has passed the rope is still there; only energy was carried.

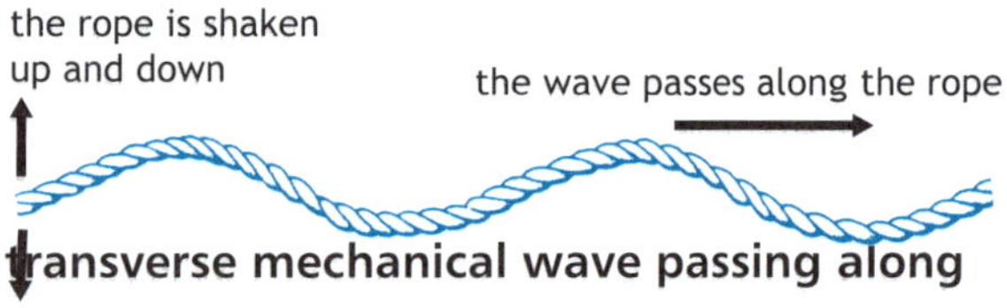

A transverse mechanical wave passing along a rope

Transverse waves

In a transverse wave, particles move at right angles to the wave. The diagram above is a transverse wave because the direction of oscillation (up and down) is at right angles to the direction of the wave's movement. The wave is travelling *along* the rope but the particles are moving *up* and *down*.

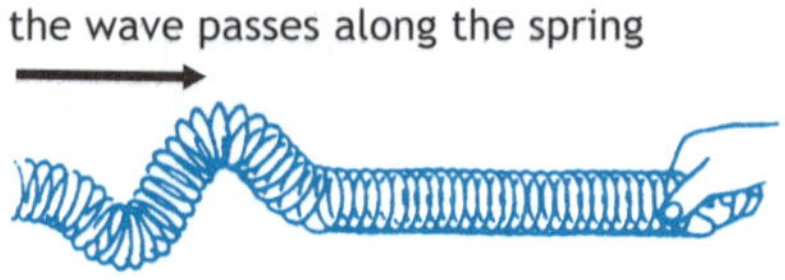

A transverse mechanical wave passing along a spring

In the diagram above energy is moving down the spring and each bit of the spring is vibrating at right angles (transversely) to the spring. Water and light waves travel transversely.

Longitudinal waves

In a longitudinal wave the particles move in the same direction as the wave. If you push a spring you can make a single pulse travel along it. As the pulse passes the coils oscillate *backwards* and *forwards* along the length of the spring. They vibrate lengthways or longitudinally. A ripple passes down he spring, but the individual coils are left where they were in the first place. Sound is a longitudinal wave.

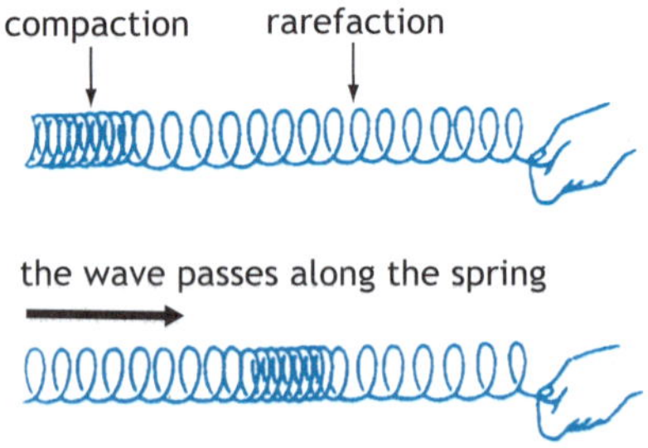

A longitudinal mechanical wave passing along a spring

As the wave passes through the spring there are sections of **compressions** where the springs are close together and areas called **rarefactions** where the springs are far apart. The compressions move along the spring.

Electromagnetic waves

An electromagnetic wave is not a disturbance of particles. It is varying electric and magnetic fields and does not need particles to pass the energy on. Electromagnetic waves travel best when there are no particles. Examples of electromagnetic waves include X-rays, gamma rays, microwaves, light waves and radio waves.

Parts of a wave

A wave passing along a rope or as a ripple through water will look like the diagram below.

Transverse wave

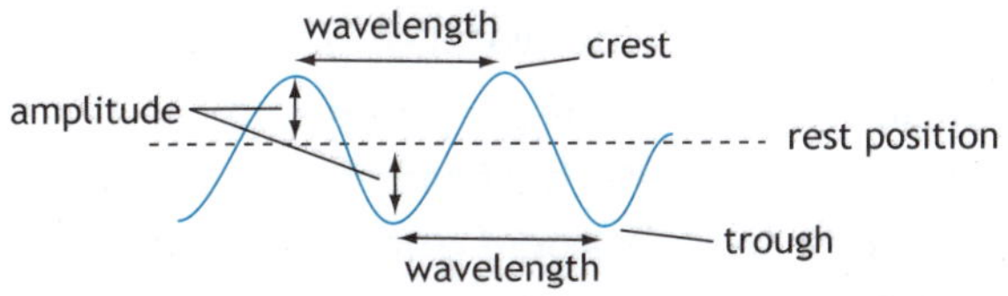

Longitudinal wave

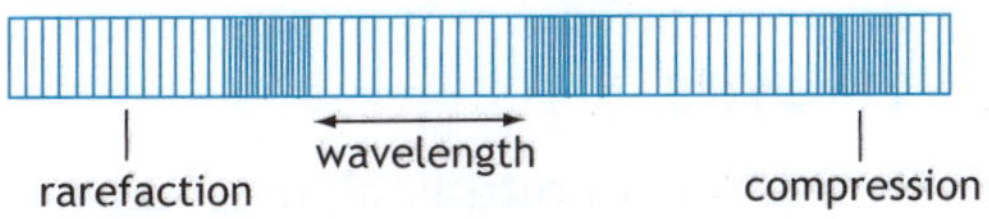

The shape and terminology of a wave

The dotted line shown represents the undisturbed water or rope. The top of the hump of water or rope is the **crest** and the bottom is the **trough**. The distance between two crests is the **wavelength** of the wave. The maximum distance that a particle moves away from the centre line is called the **amplitude** of the wave. The number of waves produced per second is called the **frequency**. Frequency is measured in **Hertz (Hz)**. If there are two waves every second then the frequency is 2 Hz. These measurements also work for longitudinal waves where the wavelength is the distance between two compressions.

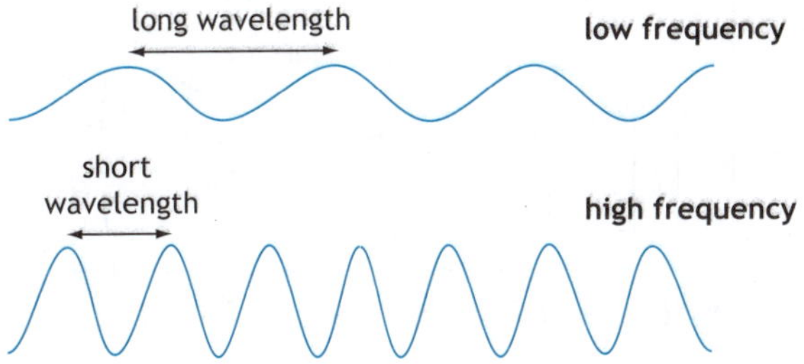

Frequency

The **speed** of a wave is the distance travelled by the wave or energy in one second. The relationship between speed, frequency and wavelength can be written as follows:

$$\underset{(m/s)}{\text{Wave speed}} = \underset{(Hz)}{\text{frequency}} \times \underset{(m)}{\text{wavelength}}$$

For Example: a wave has a frequency of 2 Hz and a wavelength of 10cm. What is the speed of the wave?

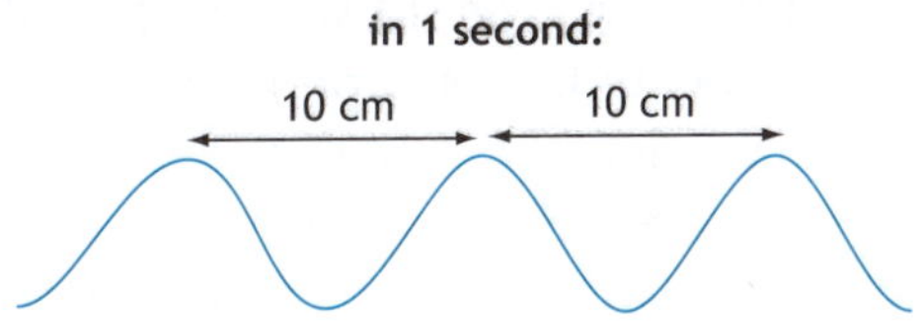

In each second, two waves are produced, each of length 10cm. So the wave travels forward 20cm in each second. The speed of the wave is 20cm/s.

The properties of waves

Reflection

When waves hit a solid object they may be reflected. The angle at with the wave hits the barrier and the angle at which the wave is reflected is the same. Light waves behave the same way when they are reflected off a mirror.

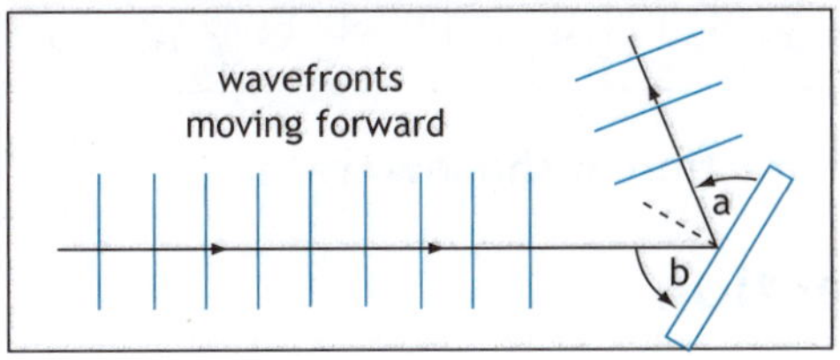

Wave reflection off a solid surface (Angle a = angle b)

If waves hit a curved surface the reflected waves converge to a point called the focus.

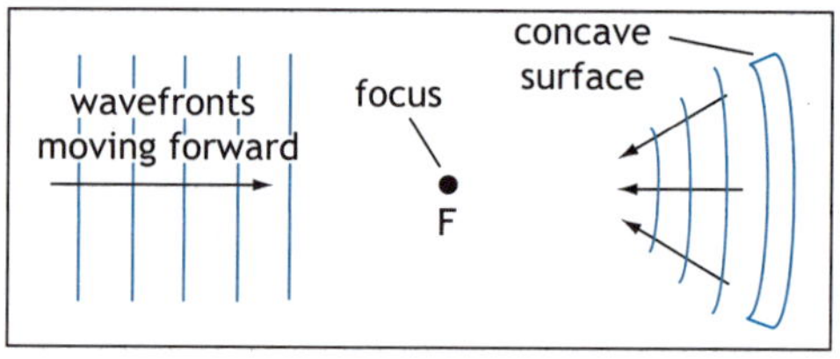

Waves reflected off a concave surface

Waves diverge off a convex surface.

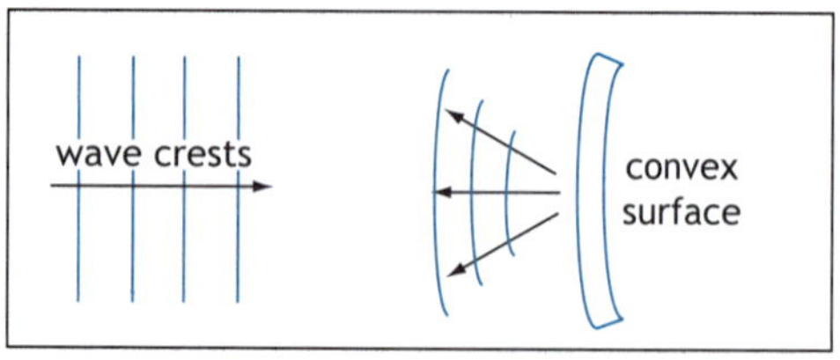

Waves reflected off a convex surface

Refraction

Waves can pass from one substance to another and are not reflected. This depends on the angle at which the wave front strikes, and the nature of the material. When this happens the speed and direction of the wave change. The speed of water waves depends on the depth of the water. As waves travel into shallower water they slow down and change direction.

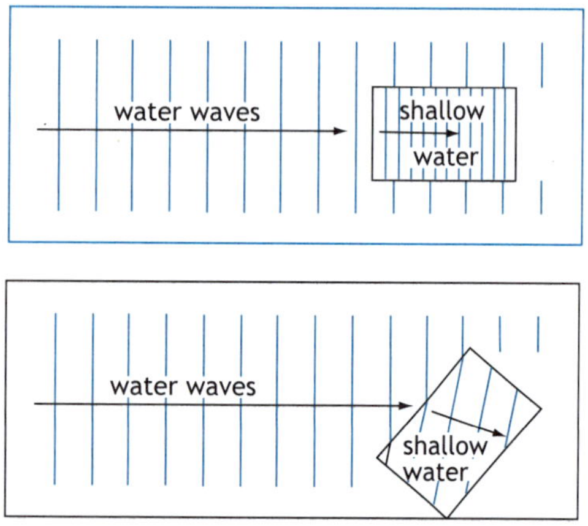

Wave refraction in shallow water

Diffraction

Waves can pass through gaps in barriers. If the gap is large the waves don't change, but if the gap is small then they spread out. This is called diffraction. Diffraction is the most obvious when the width of the gap is the same size as the wavelength.

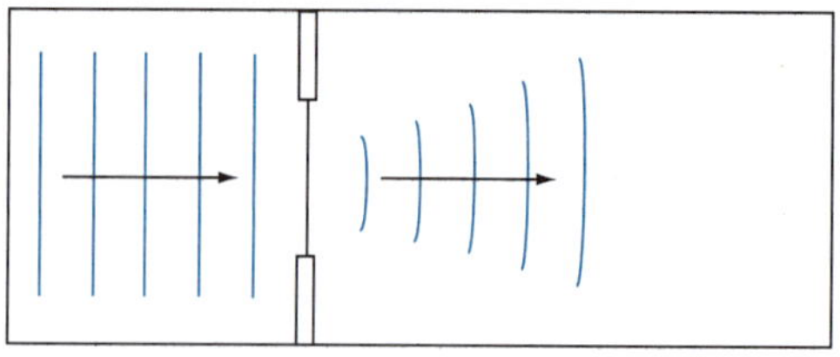

Diffraction (wide gap)

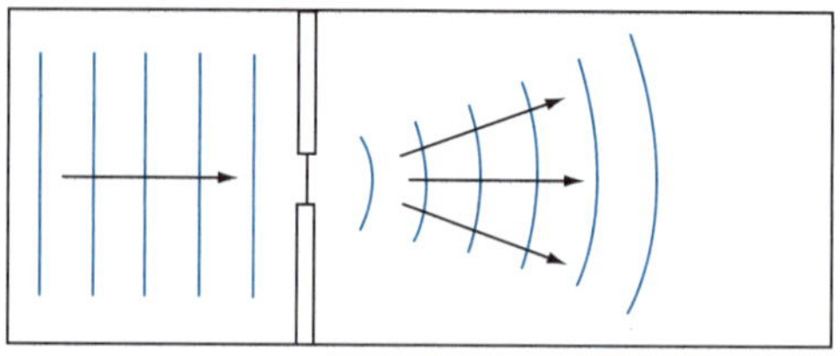

Diffraction (narrow gap)

Interference

When one wave meets another wave interference occurs.

- Destructive interference is when the crest of one wave overlaps the trough of another and they cancel out. The water appears to be flat.

Constructive interference is when two crests meet and the wave becomes twice as high or when two troughs meet and the trough becomes twice as deep.

Types of waves

Sound waves

A sound wave is a longitudinal, mechanical wave. Sound waves can bounce off surfaces. These reflected waves are used by humans and other animals to locate objects and measure distance. You can tell how far away an object is by how long it takes the echo to come back. If you shout at a wall 340 m away the sound takes one second to reach the wall and one second to return. So you hear the echo two seconds after shouting.

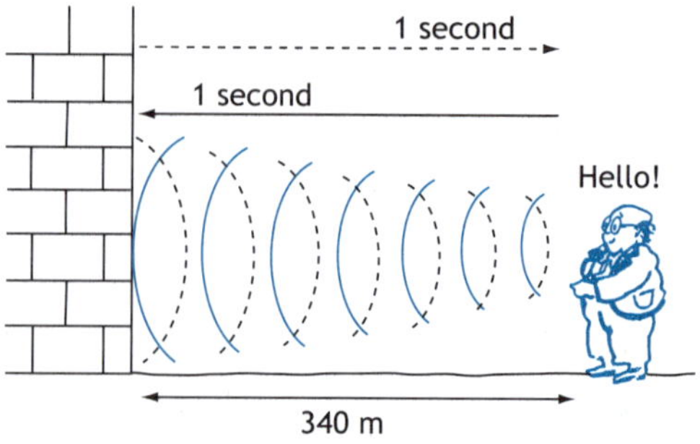

An echo

- **Sonar** — a ship uses sonar to measure the depth of the seabed and fishing boats to detect shoals of fish.
- **Echolocation** — bats and dolphins use ultra-sonic waves to navigate and to locate food.

Electromagnetic waves

These waves carry all the energy that we receive from the Sun.

- **Radio waves** travel at 300 000 000 m/s. An echo from an object 300 km away would take two milliseconds to return. Radio waves have a long wavelength and will diffract around hills. TV waves have a shorter wavelength and don't bend so much. Air traffic control systems use radar to plot the position of aircraft.
- **Infra red waves** are given off by hot objects and are used for taking temperature pictures and finding people in collapsed buildings.
- **Ultra Violet waves** are used in florescent lamps, sterilising, and in astronomy.
- **X rays** are high-energy waves used in radiography, treating cancer, finding faults in metals and examining crystals.
- **Visible light** is used for seeing, photosynthesis, photography and astronomy.
- **Gamma rays** are used to measure the thickness of metals and sterilising medical equipment.

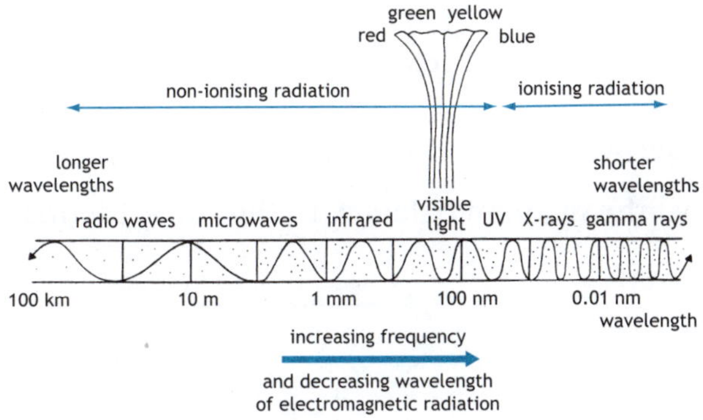

Electromagnetic waves

Lasers

A laser produces a beam of light in which all the waves are of the same frequency and exactly in step with each other. So they focus a huge amount of energy with great accuracy and can be used for cutting and surgery. A compact disc player also uses laser light.

Glossary

Amplitude
The largest distance a particle moves from the surface.

Compressions
Areas of high pressure where springs or particles are close together.

Crest
The top of a wave.

Diffraction
Waves spread out after passing through a gap.

Electromagnetic waves
These waves do not need particles to pass the energy on.

Frequency
The number of waves produced per second.

Hertz (Hz)
The unit used to measure frequency.

Interference
When one wave meets another.

Longitudinal waves
The particles oscillate along the direction in which the wave is travelling.

Mechanical waves
These waves disturb particles — the particles pass the energy on.

Rarefactions
Areas of low pressure where the springs are stretched out.

Reflection
Waves bounce back when they hit a solid object.

Refraction
Waves change speed and direction if they pass through another substance.

Transverse waves
The particles move at right angles to the wave.

Trough
The bottom of a wave.

Wavelength
The distance from one crest or one trough to the next.

Science skills

Question 1

Label the diagram below with the correct terms.

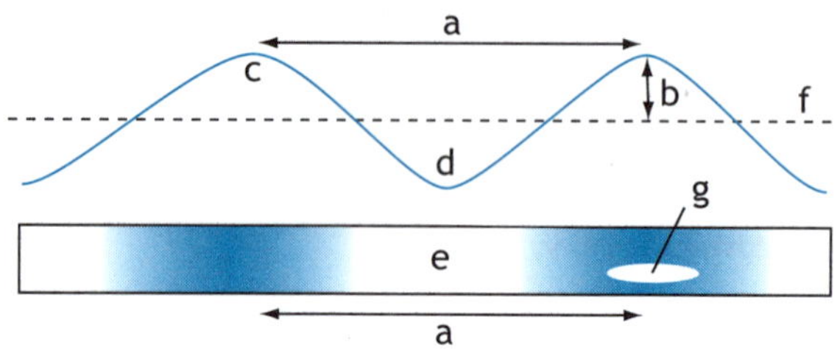

Question 2

Refer to the diagram in the text on electromagnetic radiation on page 83. Different pigments in the visible light spectrum absorb different amounts of light energy. Study the graph below and answer the following questions.

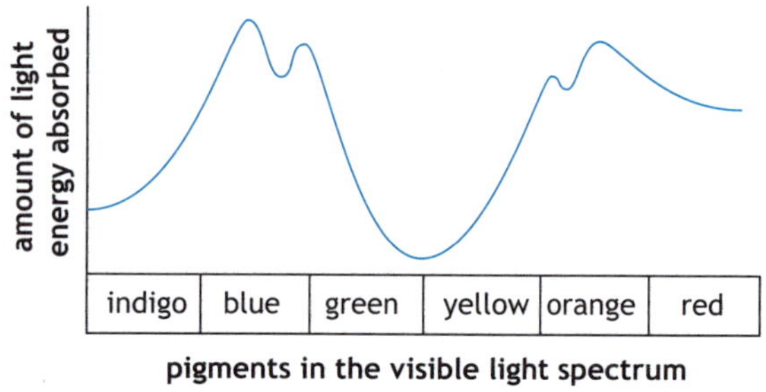

a Which colours absorb the most light energy?

b Which colours absorb the least?

c Chlorophyll is a green pigment in plants. Of which colour does it absorb very little?

Question 3

Study the following four diagrams and write the correct name of each wave property.

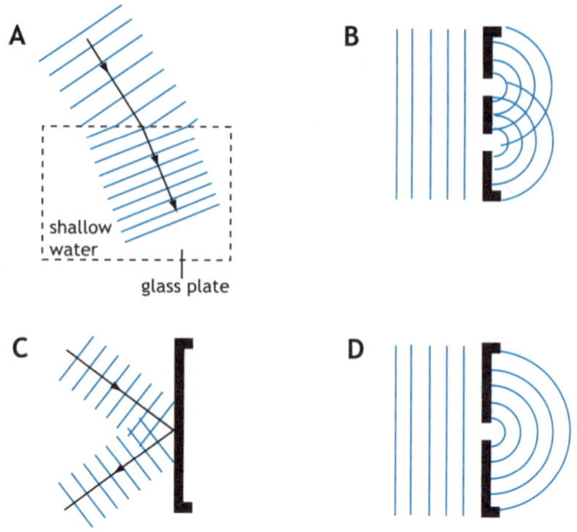

Question 4

The diagram below shows waves on a water surface.

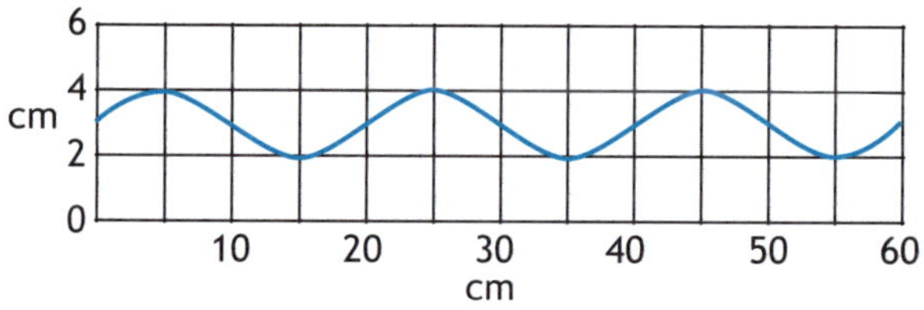

a What is the amplitude of the wave in centimetres?

b What is the wavelength of the wave?

c If the frequency at which these waves are produced is 5 Hz, what is their speed?

d What quantities and units should be in the blank spaces below?

Speed	Unit	Frequency	Unit	Wavelength	Unit
		50	Hz	2	cm
0.5	m/s		Hz	5	cm
25	m/s	20	Hz		m
500	m/s		Hz	50	m

Question 5

Which of the following frequencies is a high-pitched sound?

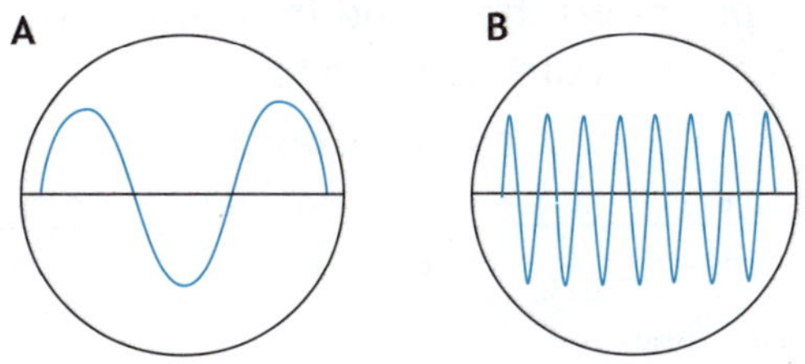

Question 6

The diagram below shows a longitudinal wave travelling along a spring with many coils.

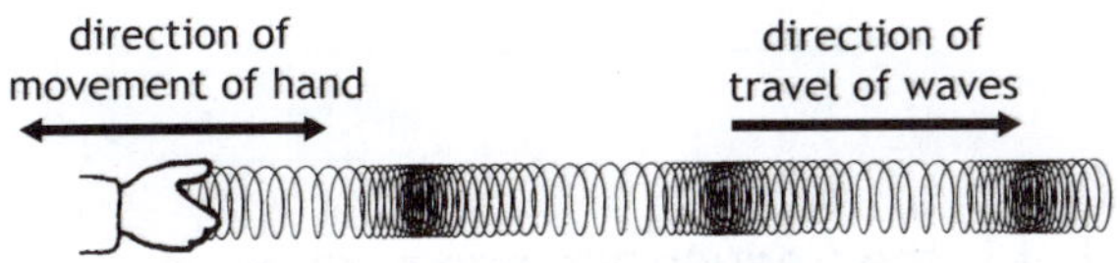

a Label two areas of compression.

b Label two areas of rarefaction.

c Label two complete wavelengths.

d Define the word 'wavelength' when it refers to a longitudinal wave.

e In the diagram below of sound waves reaching an ear, where are the rarefaction zones?

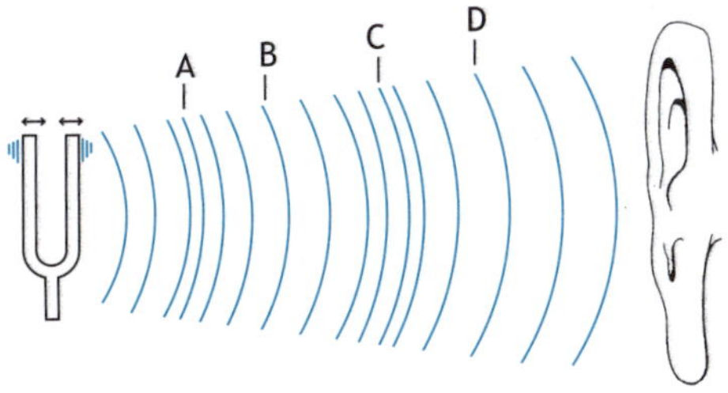

Question 7

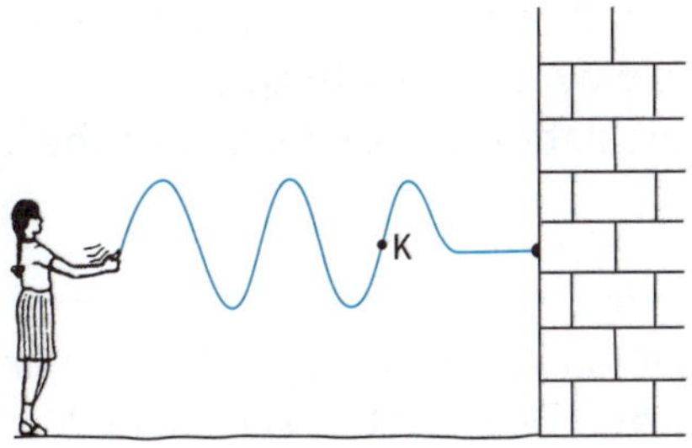

a If the girl moves her hand up and down once every second, what is the frequency of the wave produced?

b There is a knot in the rope at K. If the distance from her hand to the knot is 4m, what is the wavelength of the wave?

c How will this knot move as the wave passes it?

d What happens to the waves as they hit the wall?

e Why does the amplitude of the wave become smaller as it travels?

Question 8

In the following diagram that shows wave refraction, which angle equals angle b?

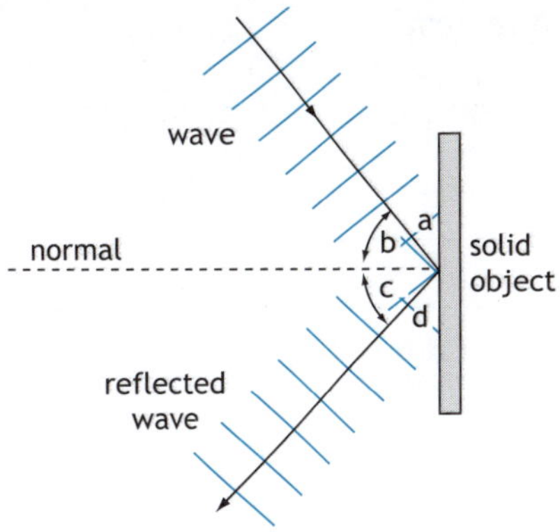

Question 9

a A series of small waves travels across a pond at 2 m/s. These waves have a frequency of 10 hertz. Calculate the wavelength of the waves.

b A water wave has a frequency of 5 hertz and a wavelength of 0.06 m. What is its speed?

c If a wave travels with a speed of 0.6m/s and has a wavelength of 0.04m. What is its frequency?

d A wave moving at 30cm/s has a frequency of 5 hertz. Calculate its wavelength.

Question 10

The following table displays the speed of sound waves through various substances.

Substance	Speed (m/s)
Air	340
Oxygen	317
Hydrogen	1 270
Water	1 432
Iron	5 000

a Through with state of matter does sound travel the slowest?

b Through which state of matter does sound travel the fastest?

c Why?

Question 11

A radio station produces radio waves of frequency 200 000 hertz and a wavelength of 1500 metres.

a What is the speed of radio waves?

b Another station produces waves at 500kHz. What is their wavelength?

Question 12

The graph shows the sound frequencies that various animals are sensitive to.

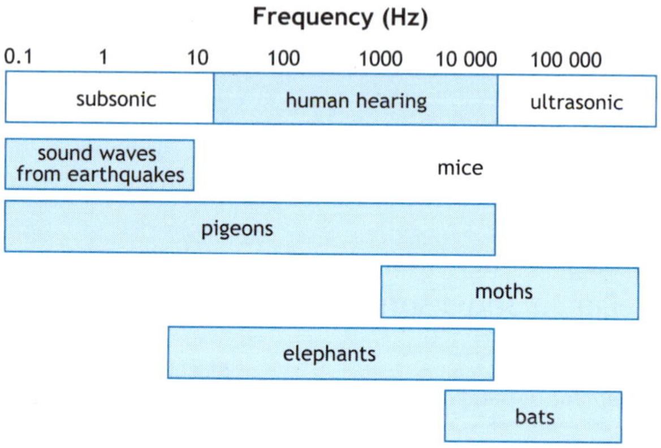

a What is the frequency range normally heard by humans?

b What is the frequency range normally heard by moths?

c Which animals could hear some or all of the sound waves from earthquakes?

d Which animals hear frequencies in the ultrasonic range?

Question 13

Study the diagram on electromagnetic waves on page 26 of the text.

Label the boxes in the following diagram with the correct word or words from the list below:

shorter wavelengths, longer wavelengths, microwaves, X-rays, increasing frequency

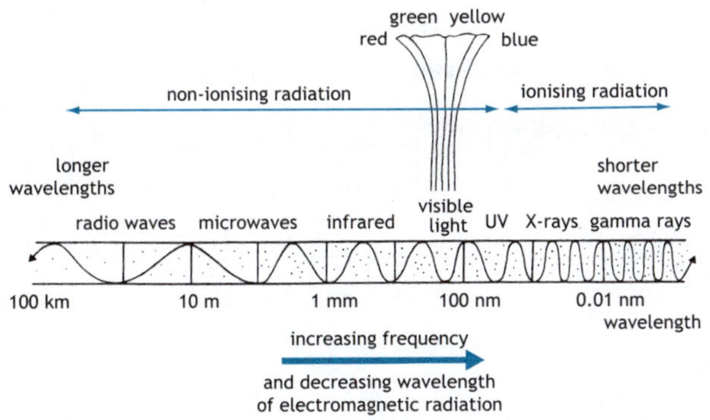

Question 14

Draw the path of the waves after they have been reflected (A) and refracted (B).

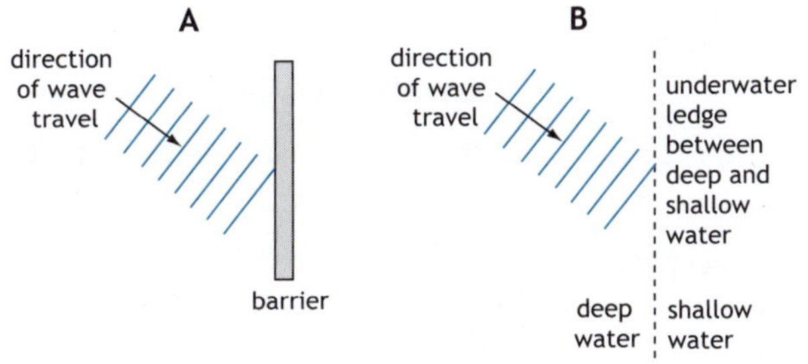

Question 15

Sound travels in air at 340 metres per second and in water at 1 432 metres per second. Calculate the wavelength of the sound emitted by:

a A bat at a frequency of 120 000 Hz.

---

b A dolphin at a frequency of 200 000 Hz.

---

c An air conditioning unit at a frequency of 5 Hz.

---

d A musical note transmitted through water at a frequency of 480 Hz.

Revision questions

Question 1

a Label the following two diagrams with the type of wave — longitudinal and transverse.

b Are these mechanical waves or electromagnetic waves?

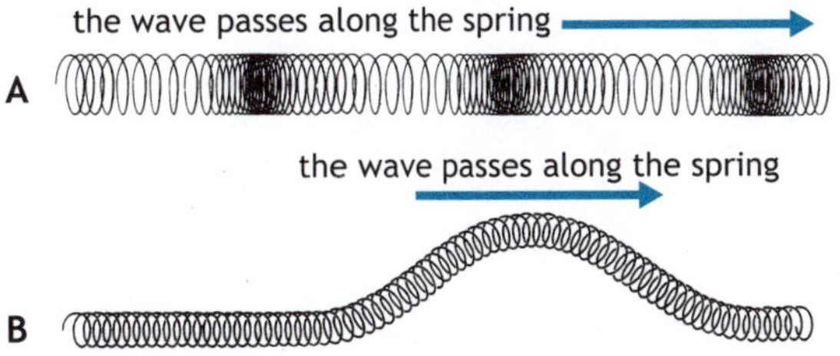

Question 2

Write the correct term for the parts of a wave to complete the table.

Term	Explanation
	Top of a wave.
	Largest distance a particle moves from the surface.
	Coils or particles are close together.
	Bottom of a wave.
	Coils or particles that are stretched out.
	Distance from one crest or trough to the next.

Question 3

Complete the sentences below with the correct words.

All waves carry ________. The speed of a wave = ________ × ________. The most important properties of waves are reflection, ________, diffraction and ________. There are two main types of wave: ________ and ________. Sound travels as a ________ wave and ________ travels as a transverse wave.

Question 4

Which two senses receive information as waves?

____

Question 5

a What do ships measure with the following system of reflecting sound waves?

____

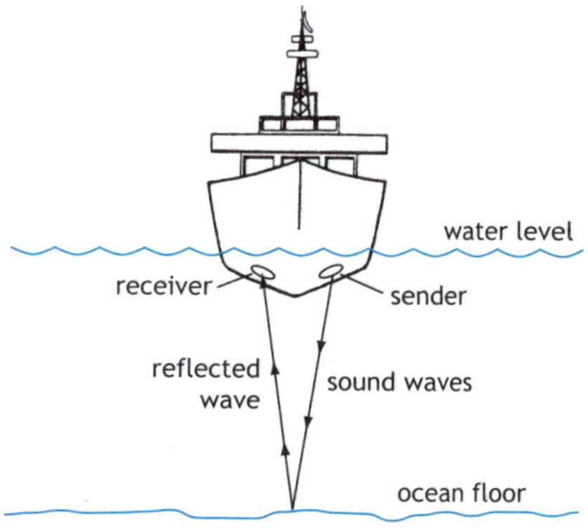

b What is this system called?

____

c Name two animals that use a similar system to find food?

____

Question 6

The diagram below shows a single wave passing along a spring. Add words to the spaces to give the correct explanation.

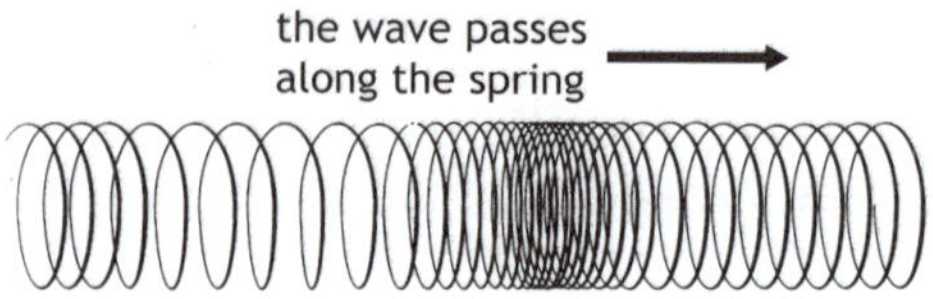

The spring as a whole ________ move along, but individual coils move ________ and ________. This is a ________ wave, in which the direction of oscillation is in the ________ direction as the movement of the wave.

Question 7

Add the following words to the correct boxes in the table below.

electromagnetic, longitudinal, transverse, mechanical

Wave type	Description
	Disturb material, or particles.
	The particles oscillate at right angles to the direction the wave is traveling.
	Does not need particles to travel.
	The particles oscillate along the direction of wave travel.

Question 8

To complete the table, write a definition for speed and frequency and the unit of measurement for each.

	Definition	Unit
Speed		
Frequency		

Question 9

Match the following electromagnetic source with its use.

X-rays	used for seeing, photosynthesis, photography and astronomy.
Gamma rays	used in radiography, treating cancer, finding faults in metals and examining crystals.
Visible rays	used for taking temperature pictures and finding people in collapsed buildings.
Infra red	used to measure the thickness of metals and sterilising medical equipment.

Question 10

A boat is moored in a boat pen and is secured by four ropes, one to each corner of the pen. What aspect of wave motion needs to be considered by the owner when tying the ropes?

____

Question 11

A student standing on a jetty notices that wave crests pass a post in the water every 7 seconds. She estimates the distance between two crests to be about 21 metres. How fast are the waves travelling?

____

Question 12

Consider what happens when the tines of a tuning fork are struck by a rubber mallet. Complete the explanation below.

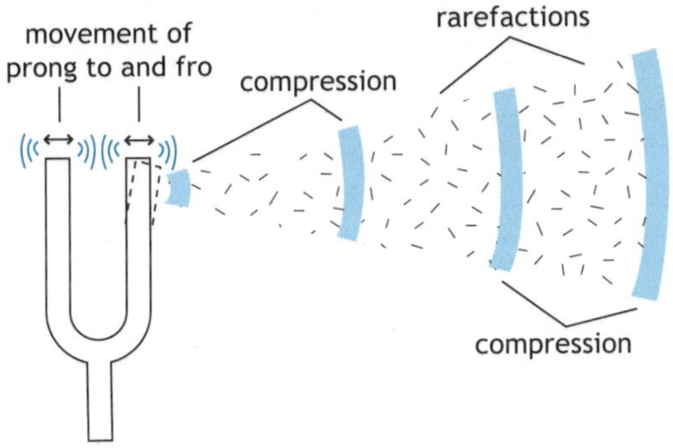

A series of __________ and __________ are set up and moves through the air. The air particles move __________ and ______ with respect to the __________ of the tines moving back and forth.

Question 13

Label the following diagrams with the terms constructive interference or destructive interference.

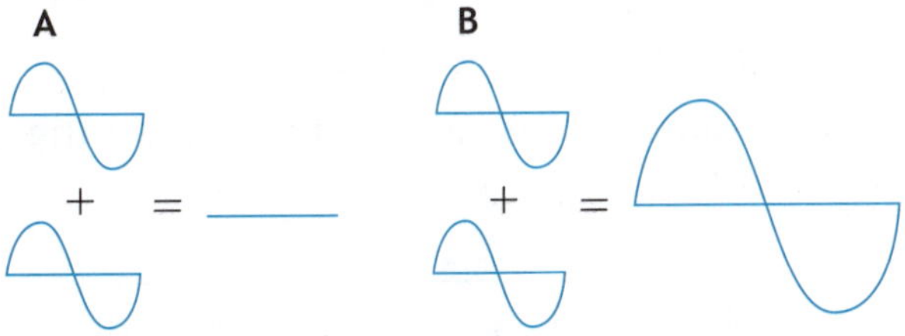

Question 14

a Which waves have the shortest wavelength radio or TV waves?

____

b Which waves (short or long) bend more easily around the hill to be received by the house?

____

c For the house in the diagram will radio or TV reception be better?

____

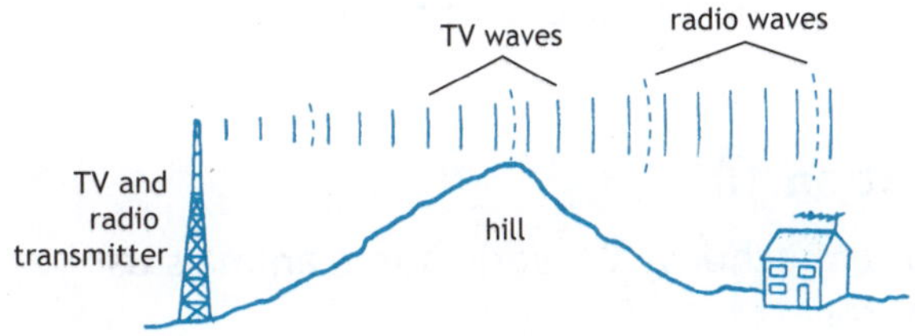

Question 15

A laser produces a beam of ______ in which all the waves are of the same __________ and exactly in step with each other. This means they can focus a huge amount of ________ with great accuracy and be used for accurate __________ and _________.

Question 16

a Label the waves in the diagram below.

b What type of wave are these?

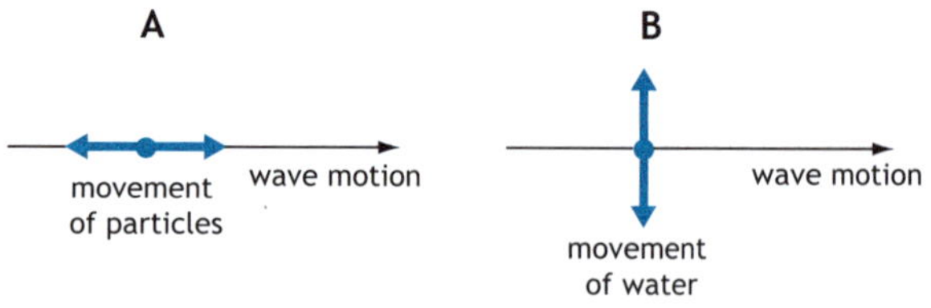

Question 17

Study the drawing below and then answer the questions.

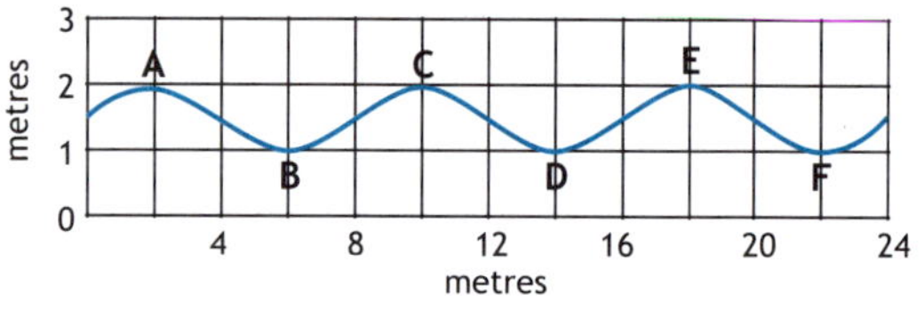

a What letters represent the crests?

b What letters represent the troughs?

c What is the wavelength of the waves?

d What is the amplitude of the waves?

Question 18

What are echoes? What do some animals use them for?

Question 19

How do the particles of water behave compared to the wave itself?

Question 20

If you stand 680 metres from a wall and knock two sticks together, how long will it take for the echo to come back?

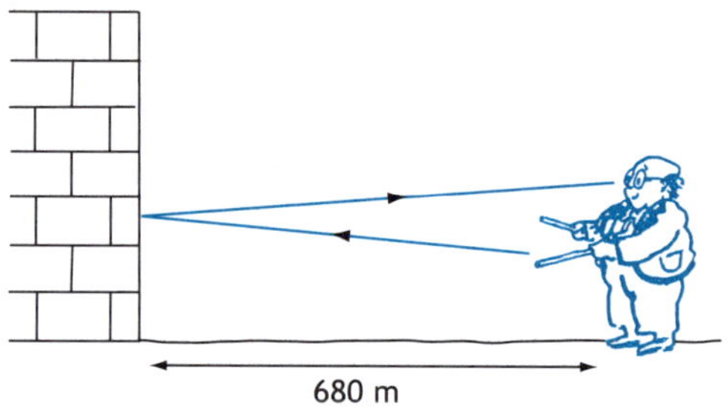

Test B

Question 1

a Indicators are special ____________. They change __________ when mixed with acids or alkalis. Universal indicator helps us to measure a substance's _____. Alkalis have a pH of _______ than 7 and acids have a pH of _______ than 7. Neutral substances like tap water have a pH _______ to 7.

b Draw up a table with two columns and list the test tubes from most acid to alkali.

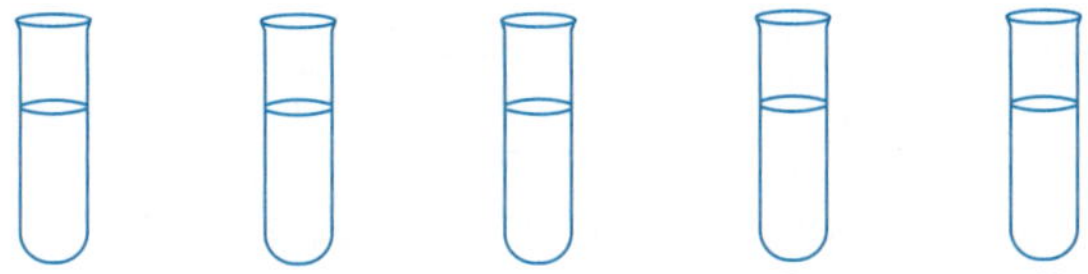

c Are the following substances likely to be acid, alkali or neutral?

Question 2

a Changes that make new substances are called _________ changes. Changes that do not make new substances are called __________ changes. Physical changes are usually easier to __________ than chemical changes. Chemical changes are produced by chemical ____________.

b List the following as physical or chemical changes:

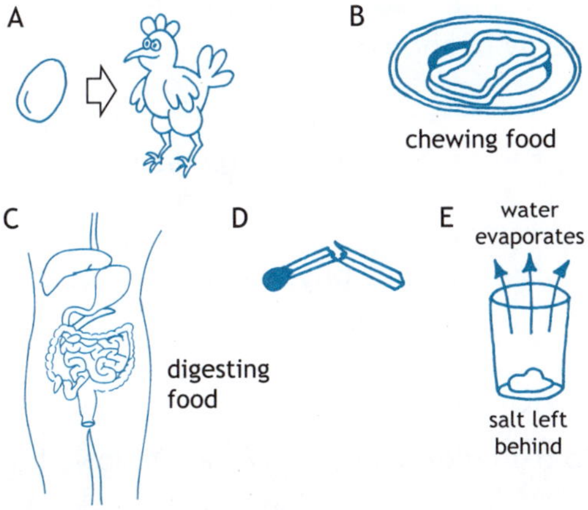

Question 3

When we mix an acid and an alkali together we can get a neutralisation reaction.

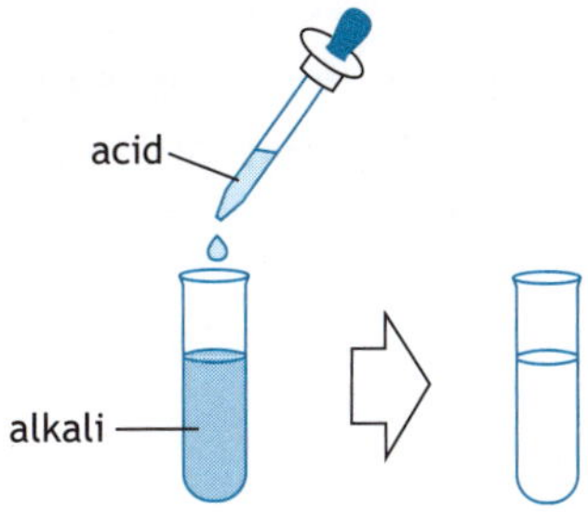

a If we add an acid to an alkali what new substances are formed?

____

b If we don't add enough acid what will be left over in this reaction?

c If we add just enough acid for the reaction what will the final pH be?

Question 4

We can change water into ice or steam. The particles in ice, water and steam are exactly the same substances but with different energy levels.

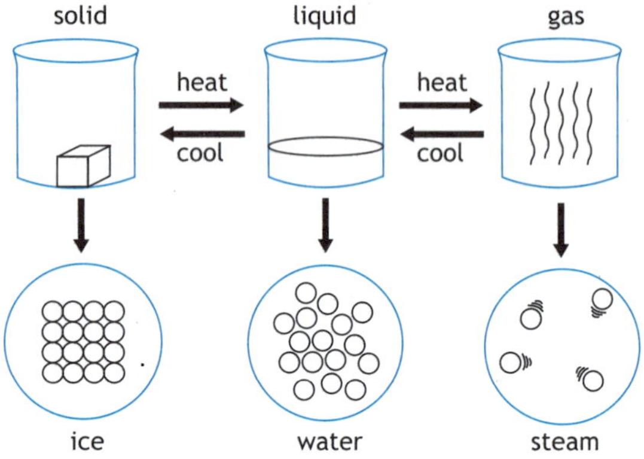

a Are any new substances formed during these energy changes?

b Are the changes easy to reverse?

c Are the changes physical or chemical?

Question 5

a Do oxidation reactions always involve burning?

b Give two examples of oxidation reactions that do not involve burning.

c Rocket engines burn fuel to drive their engines. Outside the Earth's atmosphere there is no oxygen, so where does the oxygen come from for oxidation of the fuel?

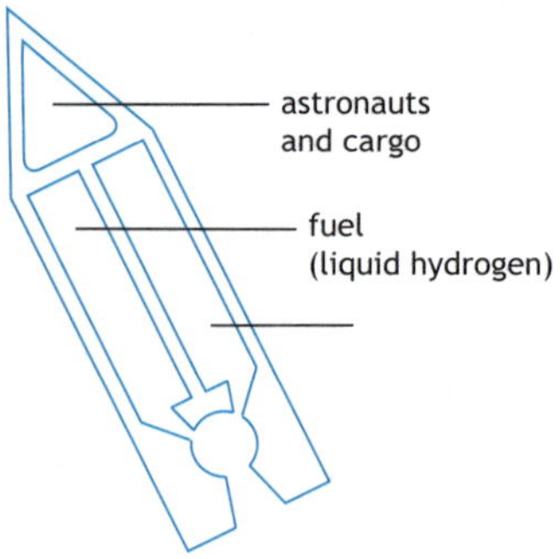

Question 6

Some substances react easily with others and some do not.

a Which acids would be most reactive?

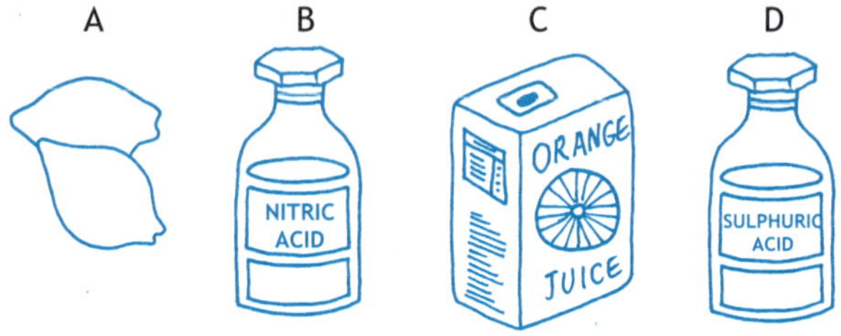

b Why do you think that reactive acids are stored in glass?

c Which of the following two metals is most reactive:

Question 7

A light year is:

a The distance light travels in a year.

b The time it takes for light to travel across the galaxy.

c 365 days compared to 366 in a leap year.

d The distance between two stars.

Question 8

A light year is often used in astronomy because:

- **a** All the information from the universe comes to us at the speed of light.
- **b** The distances involved are very large.
- **c** Astronomy is based completely on observation of light.
- **c** The speed of light is infinite and so is the universe.

Question 9

Scientists believe that the Sun formed about five billion years ago. Later the Earth formed and then life on Earth began.

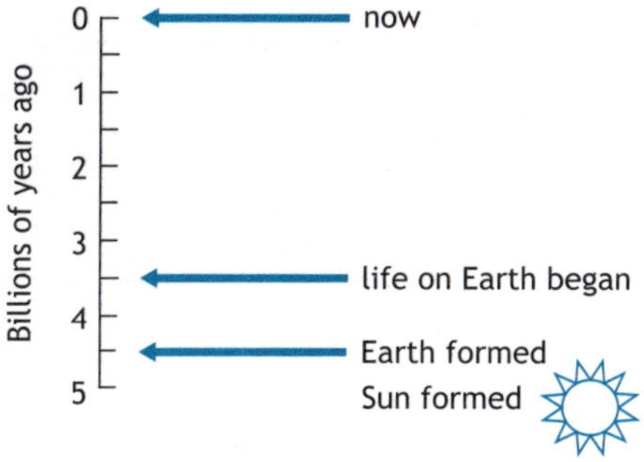

a When the Sun stops shining as it does now it will first change into a _____ _________ and later it will become a _______ _________.

b How long ago did the Earth form?

c When did life begin on Earth?

Question 10

Study the diagram below.

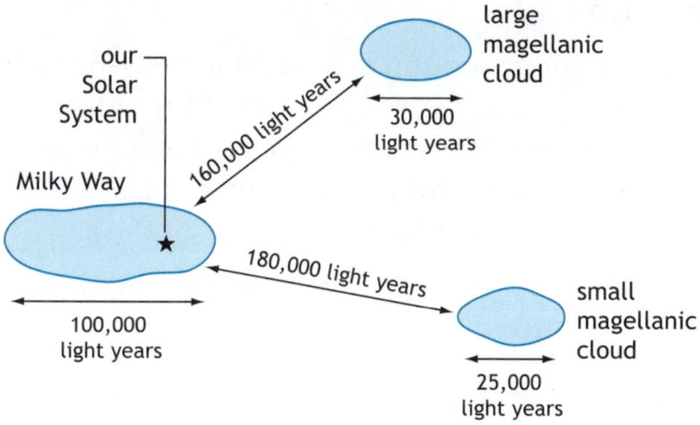

a What is the diameter of the Milky Way?

b How far is it to our nearest galaxy?

c What is the name of our nearest galaxy?

d How far is it to the Small Megellanic Cloud?

Question 11

a By what process does the Sun produce energy?

b How is this energy formed?

c What are sunspots?

d What name is given to small stars like our Sun?

Question 12

Scientists believe that all the matter in the universe at one time was concentrated in a single point. The explosion called the ______ ______ caused this matter to ________ forming the __________. They believe that the universe has been ___________ ever since.

Question 13

Some students set up the following experiment to study the Earth's position in the Solar system. The temperature before the light was turned on was 25 °C.

Data

Thermometer	Distance to the thermometer	Temperature after 15 mins (°C)
1	25	60
2	50	50
3	75	40
4	100	30

a What happened to the temperature in the thermometers when they turned the light on?

____

b How does the distance from the source of light (energy) affect temperature?

____

____

c How would the temperatures on the gaseous planets compare to those on Mercury, Venus, Earth and Mars?

____

d Based on the data do you think life could survive if the Earth moved further away from the Sun?

____

e What are some other factors besides distance from the Sun that determine the temperature on the Earth?

____

Question 14

a Besides trillions of stars and lots of gas the universe contains lots of other interesting objects. List six of these.

____

b Which of these forms after a giant star explodes?

____

c Which one provides astronomers with radio waves that help work out the structure of the universe?

____

Question 15

For each of the following, identify the issue/concern and an achievable solution.

Issue	Issue/concern	Solution
energy use		
gas emission		
waste production and disposal		
greenhouse		
ozone		

Question 16

Due to the increase in emissions of __________ ________ and _________ __________ into the atmosphere, chemical reactions with these gases and water in the atmosphere produces an acid solution of ____________ ________ or ___________. When the rain falls it has an ______ pH so it is called acid rain.

Question 17

As populations around the world increase rapidly and the demand for a higher standard of living spreads across the globe, its impact on limited resources and the fragile environment is becoming more acute.

a Identify the issues that will have an impact on human survival.

____

____

____

____

b List some strategies to overcome these issues.

Question 18

a What is the most useful thing we get from fossil fuels?

b Write down three examples of fossil fuels.

c Why are they called fossil fuels?

d What two substances do all these fuels contain?

e When fossil fuels are burned what substances do they release?

f Which environmental problem is the burning of fossil fuels making worse?

Question 19

Below is a list of freshwater pollutants. Complete the table by filling in the source and the effects of the pollutants.

Pollutant	Source	Effect
sewage		
fertilisers		
pesticides		
detergent		

Question 20

Study the graph below. In phase A there was slow controlled growth limited by disease and food supply.

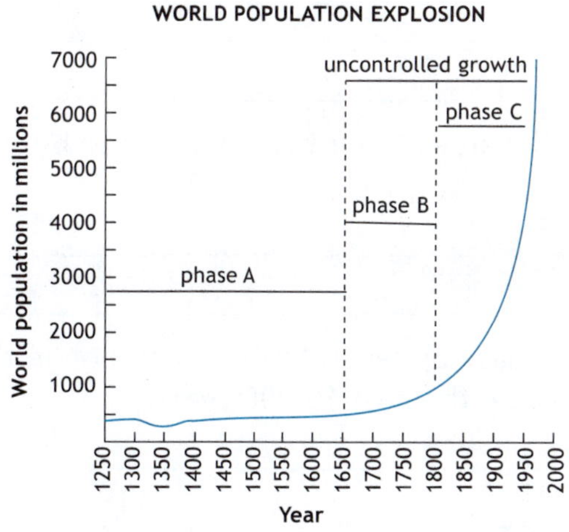

a What happened in Phase B?

b What caused the population increase in Phase C?

Question 21

Pollen was collected in core samples up to 60 000 years ago at intervals of 5 000 years.

The three main categories were:

- sclerophyll forest species
- rainforest gymnosperms
- rain forest angiosperms

Study the graphs below and then answer the questions.

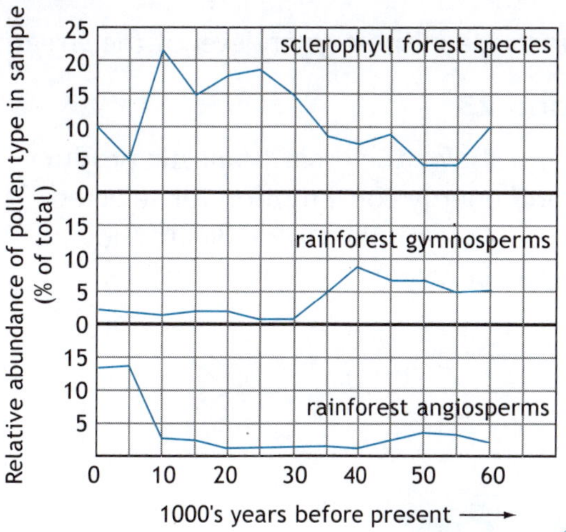

a Describe how the species composition has changed in this area over the last 50 000 years.

____

____

____

b Has the area always been dominated by rainforest? Explain.

____

____

c What has happened to the vegetation in the area over the past 10 000 years?

____

____

____

____

____

d What may be contributing to the observation in **c**?

____

____

Question 22

Eutrophication in streams is caused by:

a A decrease in the amount of nutrients available.

b An increase in the amount of nutrients available.

c A decrease in the oxygen level of the stream.

d An increase in the oxygen level of the stream.

Question 23

The graph shows the gross domestic product (GDP) and energy consumption for selected countries. Assume the GDP indicates how wealthy a country is.

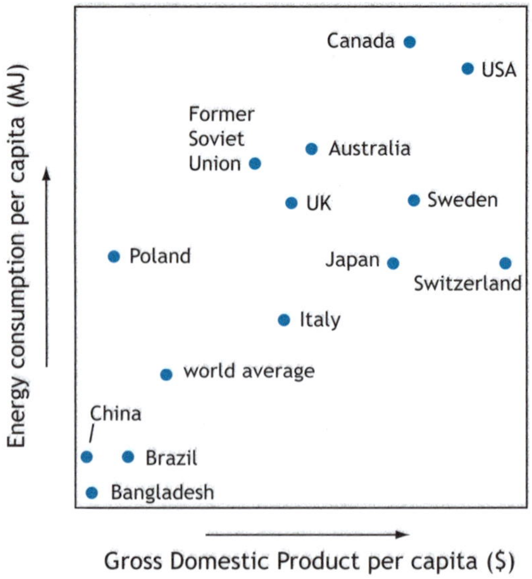

a Is there a correlation between GDP and energy consumption?

____

b Which country has the highest per capita energy consumption?

____

c Which has the lowest?

____

d What may be the likely environmental effects if countries like China and Brazil achieve a GDP similar to Australia?

____

e If you owned a business that sold goods to these countries would you be happy if these countries had a higher spending capacity?

____

____

____

____

____

Question 24

Earthquakes produces vibrations or seismic waves, called **body waves** when they are inside the Earth and **surface waves** when they reach the surface. Some body waves are transverse waves (S) and can travel through solid but not liquid, and some are longitudinal waves (P) and travel through both solid and liquid. The waves originate at the **focus** and the point on the surface directly above the focus is called the **epicentre**. These waves help scientists to work out the structure of the Earth.

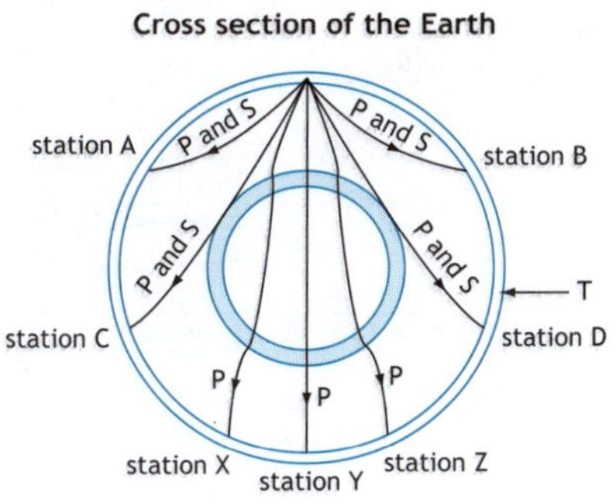

a What is layer **T** called?

b Which waves cause the most damage?

c What type of wave is an S wave?

d Through which layers of the Earth can P waves travel?

e Through which layers do S waves travel?

f Which seismograph stations receive S waves?

g Which diagram is a P wave?

A

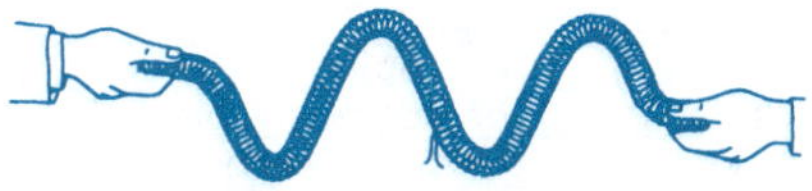

B

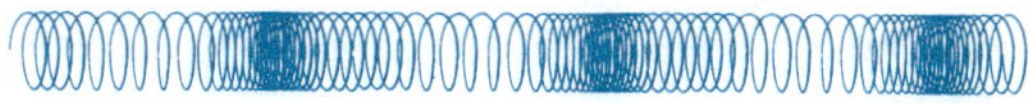

Question 25

The diagrams below show an aerial view of water waves moving towards a harbour wall.

In one there is only a single entrance, in the other there are two openings.

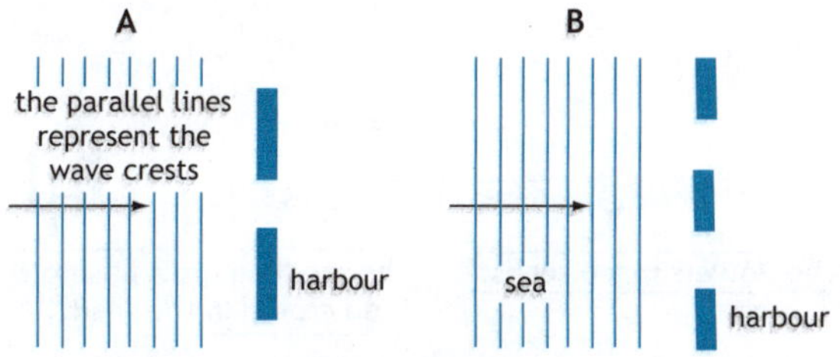

a Draw the appearance of the waves after they have passed into the harbours.

Question 26

The diagram shows some waves in a ripple tank travelling at a speed of 6cm/s.

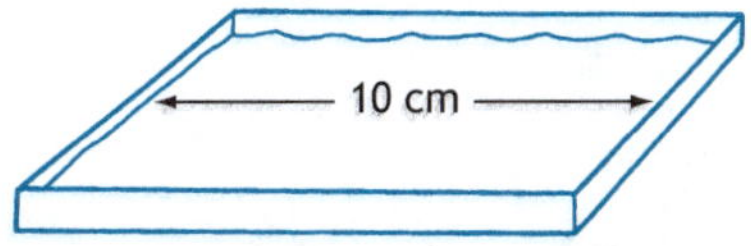

a What is the wavelength?

b What is the frequency?

Question 27

Label the diagram below with the correct terms.

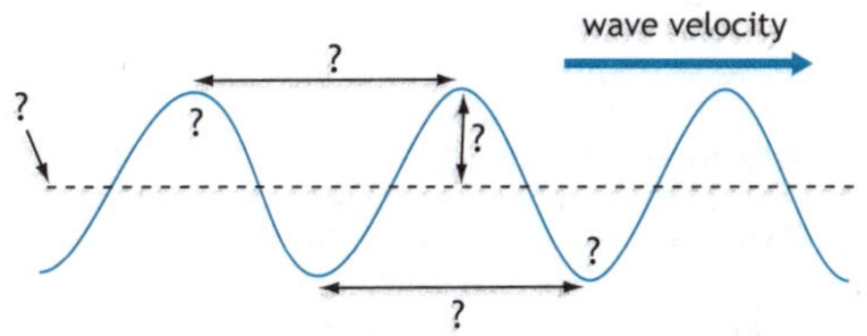

Answers

Chapter 1

Science skills page 5

1

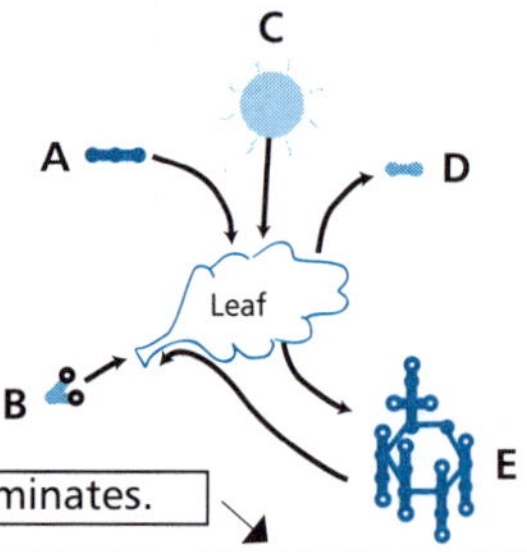

2

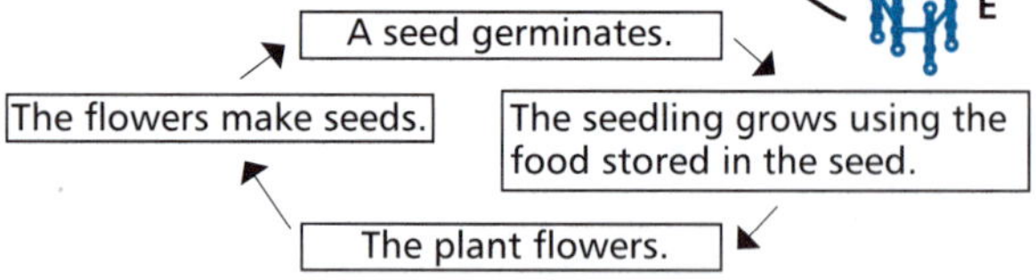

3

Meal	Healthy	Unhealthy
pizza		✔
chicken and vegetables	✔	
fish fingers and chips		✔
steak and salad	✔	
stir fry vegetables	✔	

4

Carbohydrates	cereals, potatoes, pasta
Protein	meat, eggs, fruit
Lipids	olive oil, dairy products
Fibre	vegetables, fruit

5 a When the percentage of protein is high, the young develop quickly.

b

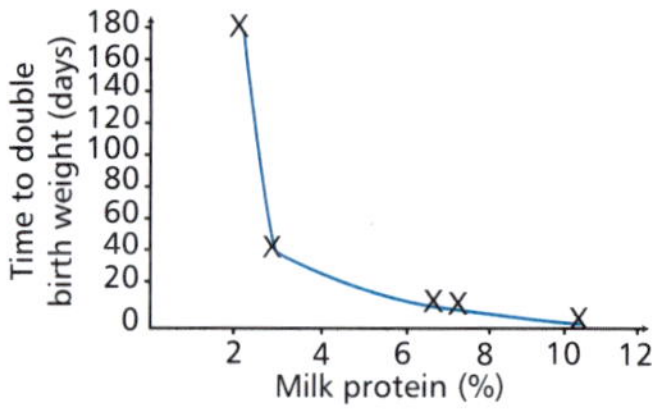

6 a chocolate shake b white tea c 0.7 grams per serve

7 a 10 000 kJ b 11 700 kJ c 10 500 kJ d 15 100 kJ

8 a 6258 kJ b 2709 kJ c student lunch A

9 a 448.35 minutes

b 448.35 min divided by 45.92 min = 9.76 times faster

10

11 a eggs

b poor eyesight

c eggs and bread

d beef

e iron

f milk, fish

g weak and brittle bones

12

Food	Carbohydrate	Protein	Fat
milk		✔	✔
carrots	✔		
cereal	✔		
eggs		✔	
butter			✔
apples	✔		
fish		✔	
pasta	✔		
chicken		✔	

13 peas, corn, tuna, orange juice, cornflakes, yoghurt and dried apricots

14

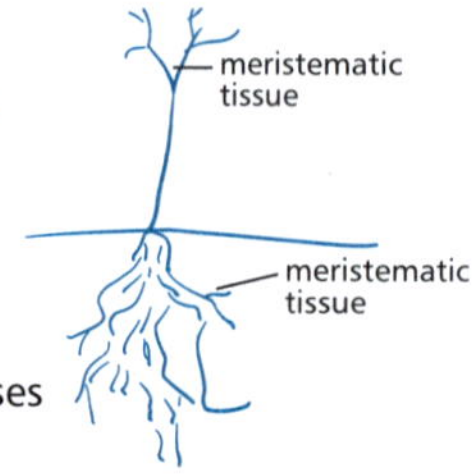

15 a When the amount of light increases the rate of photo-synthesis increases at first, then stays the same and then decreases.

b When the carbon dioxide levels go up the rate of photosynthesis increases at first and then stops increasing.

Revision questions page 8

1 A food plan that contains the correct portions of the main food groups.

2 digestive, organs, food

3 chloroplasts, carbon dioxide, water, photosynthesis, minerals

4 a Dingoes feed on animals such as rabbits and wallabies. These animals feed on grass and other green plants.

b *photo* — light
synthesis — putting together

c oxygen

5 a $6CO_2 + 6H_2O \rightarrow C_6H_{12}O_6 + 6O_2$

b $C_6H_{12}O_6 + 6O_2 \rightarrow 6CO_2 + 6H_2O$

c They are the same equation except they go in opposite directions.

6 nitrogen and potassium

7 protein — amino acids
carbohydrates — sugars
fats — triglycerides

8 a eat lots b eat moderately c eat small amounts

9 muesli, wholegrain bread, baked beans, bananas

10 a hydrogen, oxygen, carbon

b no

c nitrogen

11 a So it can photosynthesise and make food.

b

12 a

To stay healthy need	Water weed	Fish	Snail
oxygen		✔	✔
carbon dioxide	✔		
sunlight	✔		
water weed		✔	✔
fish food		✔	✔

c oxygen

13 1 Energy in sunlight

2 Energy in glucose

3 Energy in starch

4 Energy in an animal's body

5 Energy in movement

14 Sunlight energy

photosynthesis

↓

energy in glucose in the peanut leaf

↓

energy in the oil in the peanut

oxidation

↓

heat energy in the water

15 a in an animal's cells

b oxidation

c It releases energy to be used for growth, movement etc.

d Plants absorb energy from the sun and use it to produce sugars via photosynthesis. We then use these sugars in our food.

16 a lungs

b respiratory system

17 a Because they provide fibre and energy.

b butter, oils

c Fats provide fat-soluble vitamins.

d carbohydrates and fats

e person becomes fat and overweight

18 a No it is not a balanced diet.

b He could have muesli or cereal and fruit for breakfast, pasta for lunch. Fish and vegetables for dinner and fruit or a sandwich for morning tea.

19 a

Food group	Examples
carbohydrates	bread, cake, lollies
fatty foods	fries, cheese, butter
proteins	fish, meat, nuts
fibrous foods	fruit, vegetables, breads

b healthy, water, carbohydrates, fats, protein, fibre, vitamins, minerals

20 a lucerne

b nitrogen

c 110+35+9 = 154 kg

d 20 kg

e 24 kg

Chapter 2

Science skills page 16

1 4

2

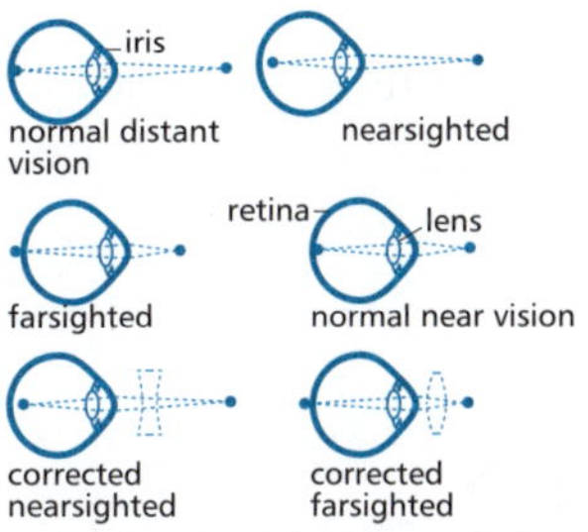

3

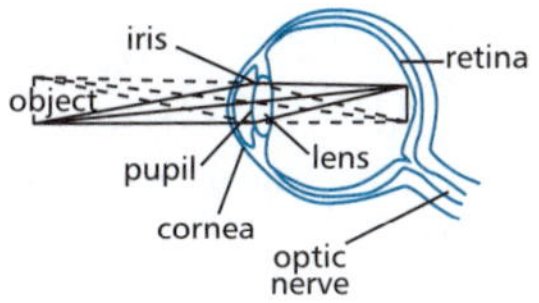

4 **a** iris — muscles control the pupil diameter
b lens — focuses objects on the retina
c optic nerve — carries nerve impulses from the retina to the brain
d pupil — diameter decreases in bright light
e retina — location of the visual receptors

5 iris = stop, retina = film, lens = lens, pupil = opening

6 The image will be upside down.

7

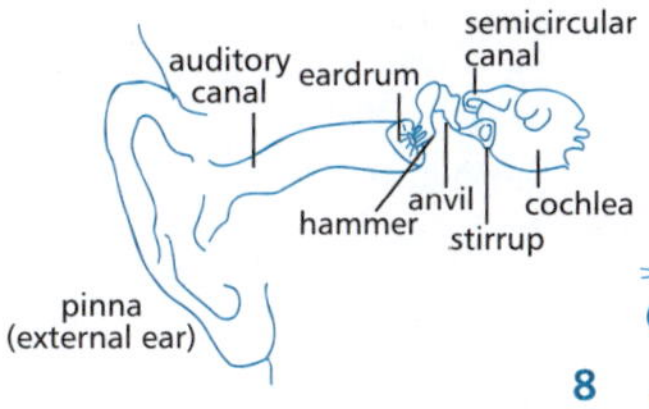

8

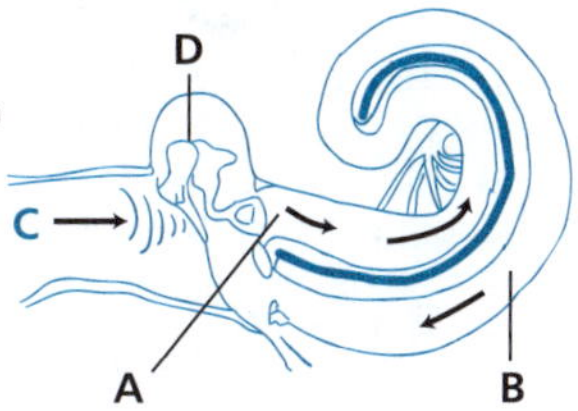

9 **a** 95 decibels (discomfort level)
b 110 dB
c 80–85 dB
d no

10

Substance	Area of tongue
sugar	1
vinegar	2
salt	2
vanilla extract	3
lemon juice	2
honey	1

11 **a**

Area tested	Prediction
cheek	C
forehead	C
upper lip	B
back of neck	A
tip of finger	B
back of hand	C
lower arm	D
upper arm	A

c Receptors are in the greatest concentration in areas that are most sensitive.

12

Animal	Senses used to find food
dingo	smell, hearing, sight
bat	hearing
owl	hearing, sight
sea eagle	sight

13 When your dominant eye is uncovered, (diagram B) the piece of wood seems to line up with your finger.
When your dominant eye is covered, (diagram A) your brain can only receive the image from the other eye.
Because your uncovered eye is not in line with your finger and the object, your finger seems to jump.

14 A bat produces bursts of high pitched sound as it flies.

The sound bounces back off any objects around it.

The bat hears these echoes through its sensitive ears.

Bats use the echoes to pinpoint their prey.

Revision questions page 19

1 **a** It controls the size of the pupil to control the amount of light admitted to the eye.
b When these images stimulate the nerves and the impulse travels to the brain, we interpret what we see as the right side up.
c It gives us depth of vision so we can estimate distance.

2 pupil, lens, focus, aqueous humour, vitreous humour

3 **a** The semicircular canals.
b The spinning forces the fluid to the ends of the canals. When the spinning stops, the fluid rushes back, causing the sensation of spinning in the opposite direction. This sensory conflict causes dizziness.

4 **a** It is in a socket of bone.
b The eye lids and lashes cover the eyeball in times of danger.
c Tears constantly wash over it, protecting it from bacterial attack.

5 Taste — a sense that detects dissolved chemicals
Taste bud — a small organ that detects dissolved chemicals
Smell — a sense that detects airborne chemicals
Olfactory membrane — a sheet of cells used to detect airborne chemicals

6 **a** contract, reduce, relax, increase
b A dim light
B bright light

7

Outer ear	Middle ear	Inner ear
pinna	hammer	cochlea
auditory canal	anvil	semicircular canals
eardrum	stirrup	

8 **a** An environmental change capable of producing a nerve impulse.
b A cell or group of cells that receive a stimulus.
c A reaction to a stimulus.
d yes
e sides
f sweet, sour, bitter, salty

9 **a** The hot hand felt cold and the warm hand felt hot.
b The reason is that the temperature we feel is relative. The hand that had been in hot water then felt a relatively cooler temperature, but to the hand that had been cold, the warm water felt much hotter.

10 **a** Lines are the same length.
b Lines are the same length.
c Both lines are the same length.
d yes

11 c

12 c

13 d

14

Sense	Stimulus
vision	chemical
hearing	physical
touch	physical
smell	chemical
taste	chemical

15 **a** The force of the vibrations reaching both ears must be the same.
b When you are positioned exactly between both speakers.

Chapter 3

Science skills page 26

1 **a** The placenta is an organ that brings the mother's blood system and that of the foetus close enough for substances to be exchanged.
b The umbilical cord.
c

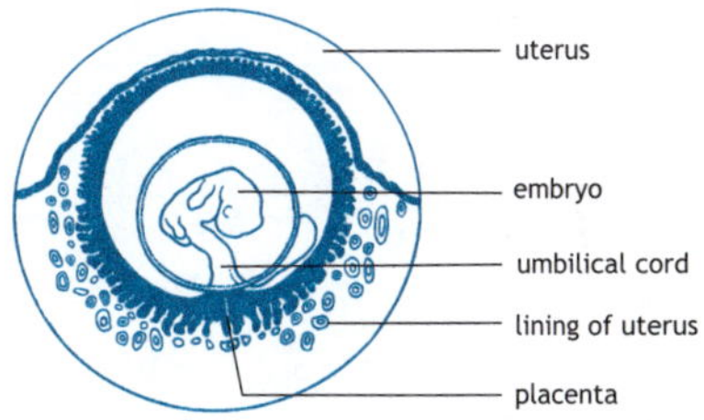

2 **Female reproductive organs** **Male reproductive organs**

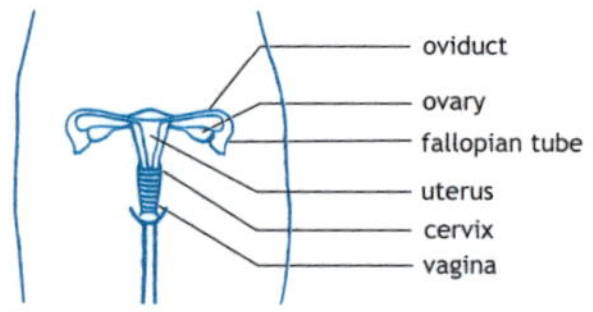

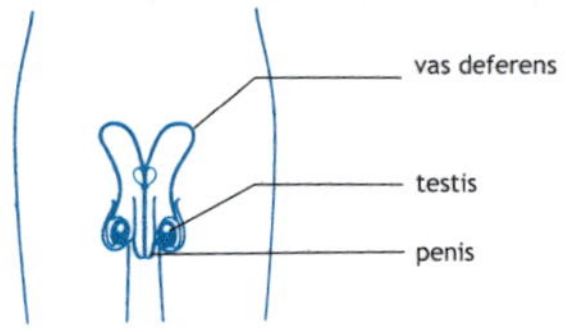

3
- During the period of bleeding the lining of the uterus is shed. (week 1)
- A new lining grows inside the uterus ready for a fertilised ovum to grow. (week 2)
- If the ovum is not fertilised it dies. (week 3)
- The lining of the uterus continues to thicken. (week 4)

4

Organ	Male	Female
fallopian tube		*
seminal vesicle	*	
prostate gland	*	
cervix		*
vas deferens	*	
ovary		*
epididymis	*	

5

Characteristic	Asexual	Sexual
involves mating		*
involves one parent	*	
produces a zygote		*
different genotype from both parents		*
involves two parents		*
involves sex cells		*
has the same genotype as the parent	*	
produces a clone	*	

6

Description	Vertebrate Group
nourish their young in a pouch	marsupials
internal fertilization; parents abandon leathery coated eggs in the soil	reptiles
eggs have a hard coat and parents nurture the young	birds
internal fertilisation and nourishment; mother produces milk	placental mammals
large numbers of eggs laid in water; development is external to the parents body	fish
reproduction linked to water, fertilisation is external, adults live on land	amphibians

7 a Monotremes.
b They lay eggs.
c Yes.
d The duck-billed platypus.

8

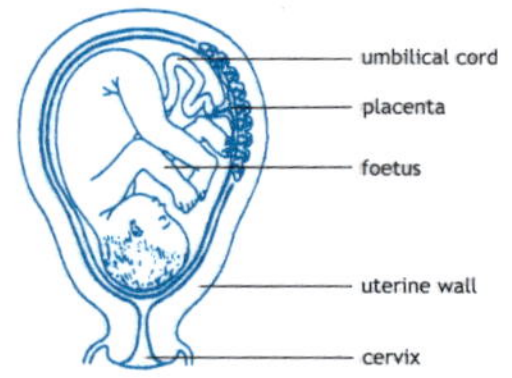

9 a Because it must prepare the lining of the uterus for implantation.
b They fall if fertilisation has not occurred.
c The lining comes away.
d At ovulation.
e The shedding of the uterine lining.
f They would remain high.

10 student 2

11 a

12

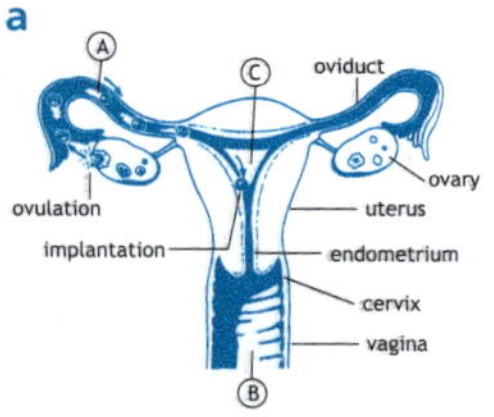

13

Function	Organ
the fertilised egg develops here	uterus
sperm is put here	vagina
male sex cells travel along here to meet the egg	oviduct
makes the ovum	ovary

14 a The placenta.
b The umbilical cord.
c Food and oxygen.
d Waste and carbon dioxide.
e No.

Revision questions page 29

1 Eggs and tadpoles would dry out if the eggs were laid on land.
2 Young mammals develop inside their mother's body and feed on her milk when they are born.
3 sex, ovum, sperm, haploid
4

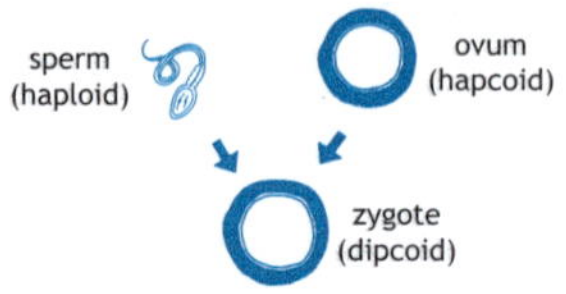

b The joining of a sperm and an ovum.
c By flicking their tails and 'swimming'.
d The penis.

5 a To produce the sperm and testosterone.
b Testosterone.
c Female.
d In the epididymis.
e The vas deferens.

6 a Hormones are chemical messengers that travel through the body in the blood until they reach their target cells.
b Oestrogen and progesterone.
c Testosterone.
d Progesterone.

7 Zygote — A single cell resulting from the fusion of a sperm and ovum nucleus.
Embryo — All development stages from two weeks after fertilisation until the eighth week.
Blastocyst — A solid ball of cells.
Foetus — All development stages from the ninth week until birth.

8 a The baby turns around so that its head points towards the cervix.
b Oxytocin causes the uterus to contract.
c The contractions force the head through the cervix and out of the mother's body.
d The contractions force the head through the cervix and the baby out of the mother's body.
e The baby takes its first breath and blood begins to circulate.
f The umbilical cord is cut and tied.
g The afterbirth is expelled

9 a

10 So that the embryo does not dry out before it has developed into a new individual.

11 Oestrogen — Promotes secondary sex characteristics in females and monitors ovulation.
Oxytocin — Produces contractions of the uterus during birth.
Progesterone — Maintains a thickened uterine lining.
Testosterone — Promotes secondary sex characteristics in males.

12 uterus, ovary, hormones

13 1. To produce ovum in the ovaries.
2. To protect the fertilised ovum as it grows to a size where it is able to survive outside the body.

14 external fertilisation

15 a Gonads produce the gametes.
b In the blood stream.
c The vas deferens.

d Seminal fluid.
e So that is will contract and assist the birth of a baby.

16 internal fertilisation

17 asexual

18 sperm, zygote, uterus, blastocyst, embryo

Chapter 4

Science skills page 36

1 a heat b acid c salt
d sugar e drying f cold

2 a 7.5 days.
b The bacterial population is not completely dead and begins to increase again.

3

Useful Microbe Sites	What They Do
compost heap	decompose dead plants and animals
bakery	make bread rise
sewage treatment	break down sewage
cow's stomach	help digest the cow's food

4 a Two days.
b The virus multiplies.
c About eight days.
d Because the symptoms appear after the peak in virus numbers.
e Antibody production.

5

Viruses	Bacteria
Measles, chicken pox, and mumps.	Food poisoning, whooping cough, and tonsillitis.
Much smaller than a single cell.	About the size of a single cell.
Takes chemicals from cells — cannot exist on their own.	Absorbs nutrients and divides and multiplies like a normal cell.
The body automatically produces antibodies but antibiotics have no effect.	The immune system attacks and destroys it. Antibiotics are effective.

6 a coughing b spit
c touch anybody d kissing

7 a acute respiratory infections
b whooping cough
c 2.1 million

8

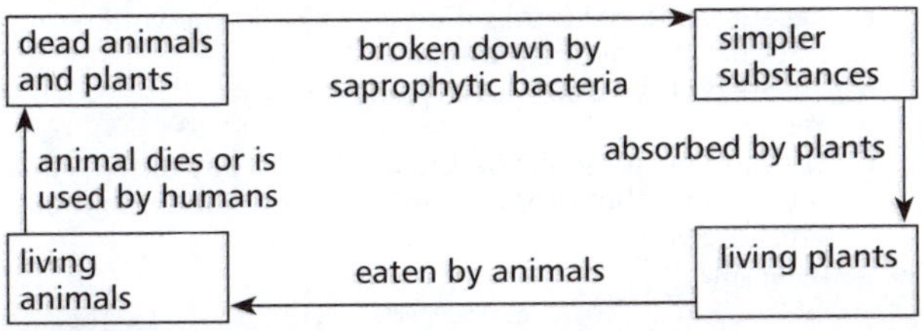

9 a Antibiotic A.
b Because it has stopped bacteria growing near it.
c As a control — to see what happens with no treatment.

10 B

Revision questions page 38

1 a A virus.
b You just have to cough, sneeze or breath on someone to transmit it.
c A sneeze spreads tiny droplets containing microbes into the air.

2

Disease	Organism
athlete's foot	fungus
chicken pox	virus
measles	virus
malaria	parasite
salmonella poisoning	bacterium
tuberculosis	bacterium
AIDS	virus

3 Through the mouth, the skin and the nose.

4 viruses, bacteria, lymphocytes, antibodies

5 Antiseptic — A substance that stops the growth of disease-causing organisms in wounds.
Antibiotic — A substance that stops certain bacteria from reproducing.
Sterile — Something that has no viruses, bacteria or other organisms on it.
Disinfectant — Something that kills disease-forming organisms, e.g. heat and sunlight.

6 a Viruses and bacteria spread in similar ways (droplets, contact and vectors). Vaccines have been produced for both viral and bacterial infections.
b No, influenza is a virus.
c An antibody is a protein that locks onto a specific foreign substance. Antibiotics kill bacteria.

7

Disease	Symptoms	Transmission
ring worm	raised, scabby circle	contact with infected skin
chicken pox	itchy spots on the skin	close contact
AIDS	poor immune system	contact with body fluids
salmonella	diarrhoea and feeling sick	eating infected food
measles	covered in small red spots	close contact

8 a
b Phagocytosis = B
Antibody production = A

9

Term	Description
vector	A living thing that spreads parasites or pathogens.
virus	Genetic material that can only grow inside a living cell.
antigen	Foreign substances on the surface of a pathogen.
antibody	A protein that locks onto a specific foreign substance.
parasite	An organism that benefits at the expense of the host.
pathogen	A disease-producing organism.

10 d

11 d

12 b

13

Pathogen	Disease 1	Disease 2
fungus	athlete's foot	ring worm
virus	AIDS	mumps
parasite	malaria	amoebic dysentery
bacterium	tuberculosis	tonsillitis

14 immunisation, dead, weakened, antibodies

15 Immunisation — Injection of dead or weakened pathogens and the person does not get the disease.
Immune system — Produces substances specifically for attacking invading organisms.
Immunity — Once the body has been infected it 'remembers'.

16 a Allergies are the result of the immune system working too hard and harming the body as well.
b Histamines.
c Hay fever.

17 lymphatic system, lymphocytes, lymph nodes

18 A = pathogen enters the body
B = microbes multiply fast
C = microbes are being destroyed by the body
D = microbes are all dead and the person regains health

19 Stage 4.

20 a White blood cells.
b They engulf and digest pathogens or they produce antibodies.
c No, you are immune.

Test A page 42

1 energy, sugar, starch, photosynthesis, carbohydrates

2 a

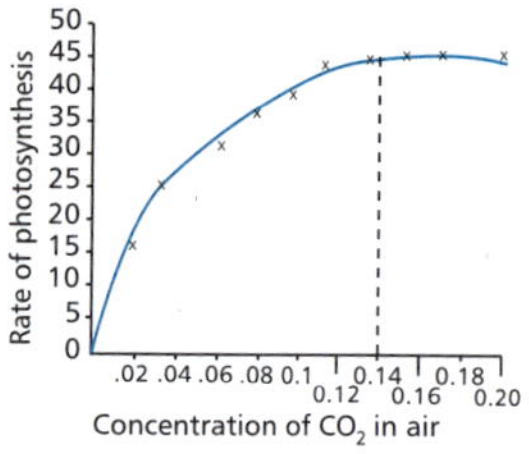

2 b about 26%

c Because this is the fastest rate for photosynthesis.

d Yes it is, by pumping carbon dioxide into a closed glasshouse.

a 0.14% of air because this is the level that photosynthesis goes the fastest.

b So they can be sold in a shorter period of time, while maintaining constant stock levels.

3 a

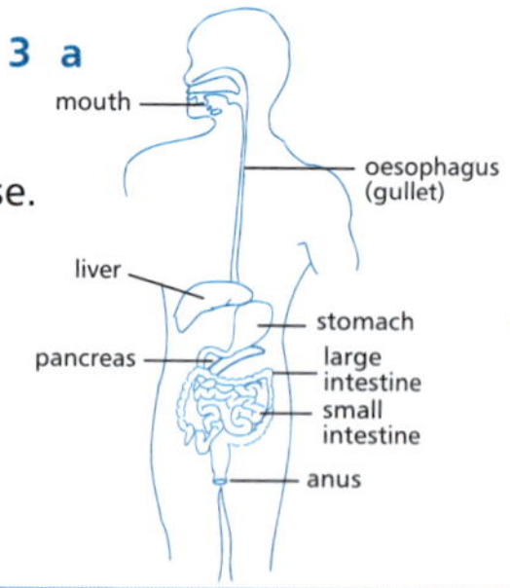

3 b

Organ	What it does
mouth	breaks up food and digests starch
oesophagus	transports food to the stomach
liver	juice (bile) splits up fats into tiny drops
stomach	breaks up food and digests protein
small intestine	digestion and absorption of sugars, proteins & fats
pancreas	produces enzymes that digest, starch, protein and fat
large intestine	water absorption and compaction
anus	expulsion of waste material

4 Starch breaks down into glucose molecules.
Protein breaks down into amino acids.
Fat breaks down into glycerol and fatty acids.

5 a Enzymes break food down into small molecules in your digestive system.
You absorb the small molecules into your blood stream through the walls of the small intestine.
Your blood transports small molecules around your body.
They pass through the walls of the capillaries into your cells where they are used.

b

Respiring cells take in:	Respiring cells give out:
glucose	carbon dioxide
oxygen	other waste & water

6 a C

b A

c Group B is very high in fat, salt and sugar.

7 a (2) Water and minerals pass into the roots.
(3) Oxygen passes into the air from the leaf.

(4) Carbon dioxide passes into the leaf from the air.
(1) Energy from sunlight is absorbed by the chlorophyll in the leaves.

b

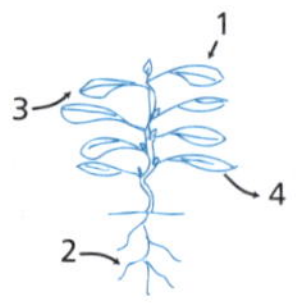

8 When the energy from a food (B) the energy needs of the body, a persons weight remains the same. They do not put on weight or take it off.
When the energy from food (A) the energy needs of the body the result is person II. When the energy from food (C) the energy needs of the body this is the result person I.

9 a

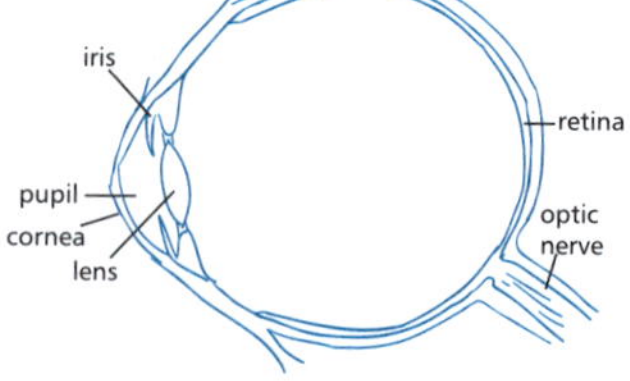

b cornea, aqueous humor, pupil, lens, vitreous humor, lens, retina, light, optic

10 a

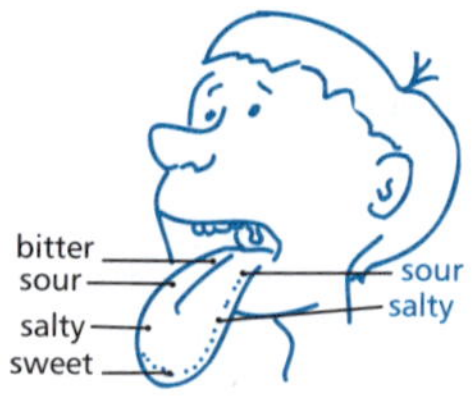

chemicals, nerves, messages, smell

11 a

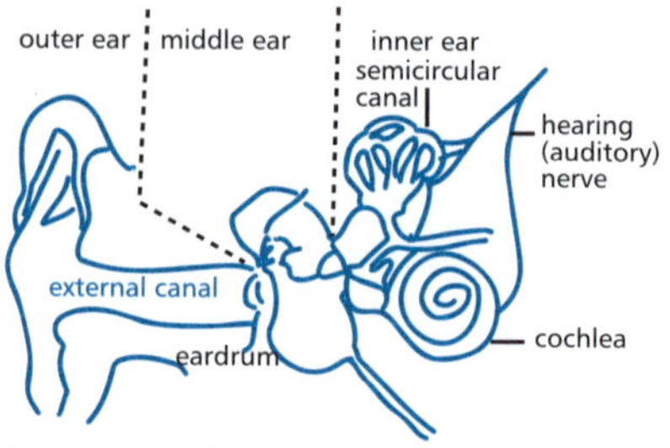

b bats and dolphins

12 a 28 days

b To prepare the uterus for the fertilised egg.

c

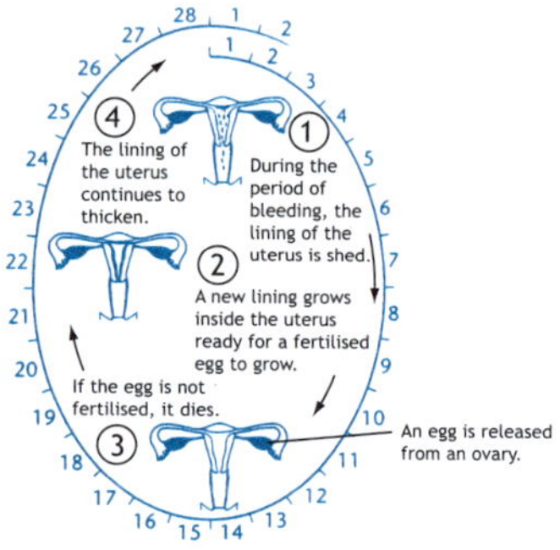

13 a Internal fertilisation occurs inside the body and external occurs outside the body.

b External because the male spreads the semen over the eggs, water providing the medium.

c Because external fertilisation is risky and the young are not nurtured by the parents.

d So the embryo does not dry out.

e Yes

f They are nurtured inside the mother's body until birth and then the mother produces milk as food. She also looks after the young.

14 FRONT VIEW

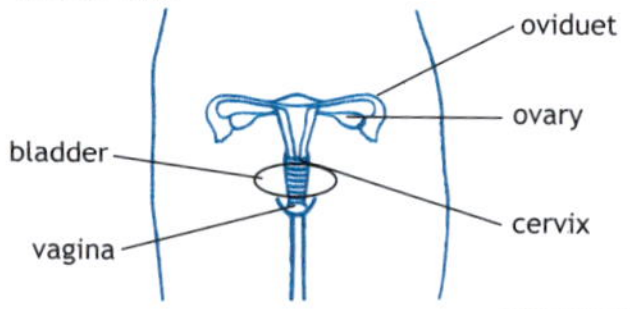

SIDE VIEW

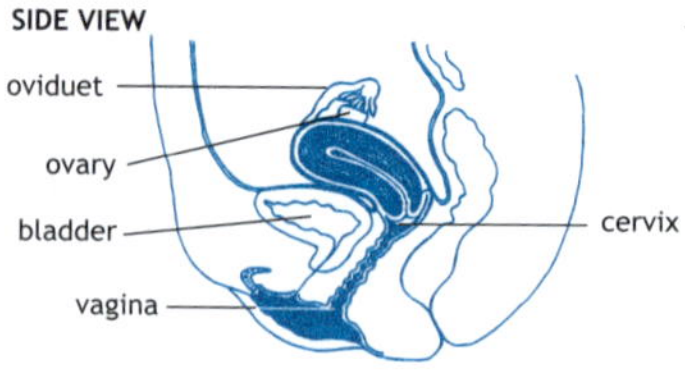

15 a. and b.

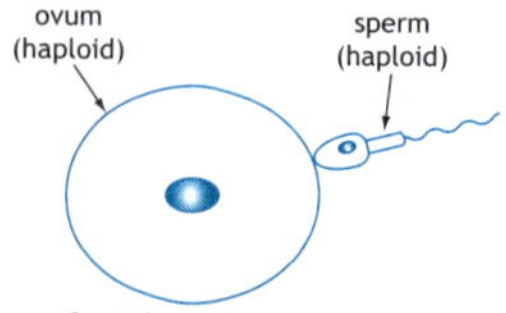

c fertilisation

d zygote
e diploid

16 a 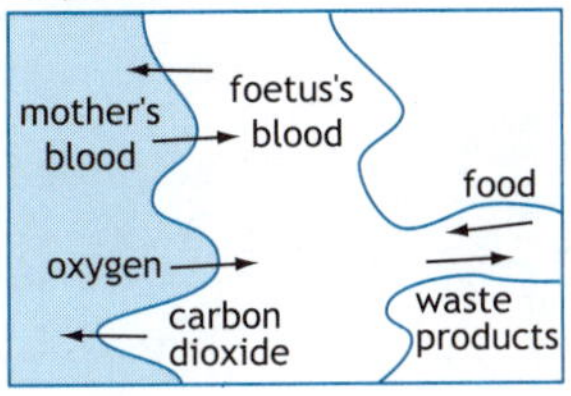

b The placenta.
c The umbilical cord.

17 asexual — There is only one parent or individual involved.

18 Zygote — A single cell resulting from the union of a sperm and an ovum.
Foetus — All developmental stages from the ninth week to the birth.
Blastocyst — The ball of cells with surface layer and inner cell mass.
Embryo — All development stages from two weeks until the end of the eighth week.

19

Structure	Function
testes	sperm and testosterone production
epididymis	maturation of sperm
vas deferens	carries sperm to the urethra
seminal vesicles	produces seminal fluid
prostate gland	secretes fluid that activates the sperm
penis	delivers the sperm
seminiferous tubules	site of sperm production

20 a hormone A = oestrogen
hormone B = progesterone
b oestrogen
c testosterone

21 a There would be no way to treat some bacterial diseases so people would become very sick and possibly die.
b Misuse by the population and over prescription by doctors.
c We must be very careful to use all the antibiotics that are prescribed. Doctors must only prescribe them if they are sure they will have an effect.

22 a By eating contaminated food.
b Good hygiene and care in the storage and preparation of food.
c The chicken salad and the apple pie.
d The chicken salad is probably the initial infection and the apple pie may have been cut up with the same knife or on the same board as the chicken.

23 a A = bacterium B = virus.
b The bacterium.

24

Disease	Virus	Bacteria
warts	*	
pimples		*
mumps	*	
diarrhoea		*
typhus		*
AIDS	*	
middle ear infection		*
influenza	*	

25 a A breakdown in the body's stable state.
b Pathogens — bacteria, viruses, protozoon and fungi — make you sick.
c Skin, the mucus membranes, the acid in the stomach and the immune system

Chapter 5

Science skills page 52

1 a Burning the wood and the match produce gases such as CO_2.
b The match and the wood turning black by burning. This is a chemical reaction called oxidation.

2 1. A mixture of petrol vapour and air is sucked in from the carburettor
2. The piston moves up to squeeze the mixture of petrol and air which gets hotter.
3. A spark from the spark plug ignites the mixture of air and petrol vapour.
4. The exhaust valve opens and the waste gases and heat are expelled.

3 a Nails 1 and 3.
b The oil stops more oxygen becoming dissolved in the water.
c Nail 2.
d Air and water.
e Yes, because Nail 2 rusted the most.

4 a The statue in Sydney.
b Because the air in Sydney will have more salt in it than that in Alice Springs.

5 a B b B

6 a Bubbles of gas are being produced. b Hydrogen.
c When substances are hot they react much faster than when they are cold. d Magnesium.
e Because copper is not very reactive.

7 a

pH	Substance
Equal to 7	Neutral
Less than 7	Acidic
Greater than seven	Alkaline

b pH = 3 c pH = 13

8

liquid	Universal indicator colour	
A	Dark blue	(5)
B	Red	(1)
C	Yellow	(3)
D	Green	(4)
E	Orange	(2)

9 a

Product	Colour	Acid	Base
toothpaste	bluish		*
bleach	Dark blue		*
shampoo	Pale yellow	*	
Dish detergent	Yellowy green	*	
Lemon juice	orange	*	

b They are all corrosive.

10 a The magnesium reaction.
b The magnesium reacted the fastest.
c The iron reacted the slowest.
d Magnesium, zinc, then iron.

11 a Water temperature, substance concentration, surface area and agitation.
b Volume and time.
c Use two tablets, grind them up, put them in hot water and stir.

12 a The cork blew off! b An acid and an alkali. c A gas.

13 a The iron nail in jar 3 rusted.
b Rust weakens the metal and makes nuts and bolts hard to loosen.
c It did not rust.
d Galvanising or coating metals will help to reduce rust.

14

Colour					orange	yellow					dark blue			purple	
pH	0	1	2	3	4	5	6	7	8	9	10	11	12	13	14

ACID (pH 1–6) ALKALI (pH 8–14)

Revision questions page 56

1 a Petrol. b In the engine cylinder. c Air.
2 Fuel, petrol, combustion, power.
3 b
4 They would corrode in the salt water and go green.

5 They are made of iron and exhaust contains both water and air. Stainless steel does not rust but it is more expensive.
6 a Iron and steel.
  b Because they are more reactive.
7 a Coating iron with zinc.
  b Zinc (not very reactive) is slow to corrode so it protects the iron for many years.
8 a Sodium is stored in oil to stop it reacting with water vapour in the air.
  b Hydrogen and sodium hydroxide.
  c Sodium (most reactive), magnesium, then copper.
9 a Dilute hydrochloric acid and magnesium.
  b Hydrogen and magnesium chloride.
  c A chemical reaction.
10 a new, chemical.
  b sulphuric, reactive, vinegar, lemon juice.
11 They repel water from the skin because water is not able to mix with the oils in the cream.
12 a The wood reacts with oxygen from the air.
  b Water vapour, carbon dioxide and charcoal.
13 For a bee sting add an alkali and for a wasp sting an acid.
14 Moisture in the air combines with carbon dioxide to form a weak acid called carbonic acid, which reacts with limestone when it rains.
15 a Iron and zinc.
  b Because the zinc is less reactive and corrodes much slower, so protecting the iron.
16 If the tin coating is damaged then the exposed iron will corrode, spoiling the food in the can.
17 The chain is oiled.
The can has a layer of tin covering the steel.
The dish drainer is coated in plastic.
The garden furniture is painted.
18 Oxidation, air, water, barrier, corrode.
19 a Use an indicator. b Neutralisation.
  c Salt and water. d Indigestion tablets.
20 a The pH gets lower.
  b Because the bacteria in the mouth produce acid from the sweet material on teeth.
  c The acid may eat through the enamel.
  d Because toothpaste is alkaline and neutralises the acid.

Chapter 6

Science skills page 64

1 E, I, C, B, F, H, A, D, G
2

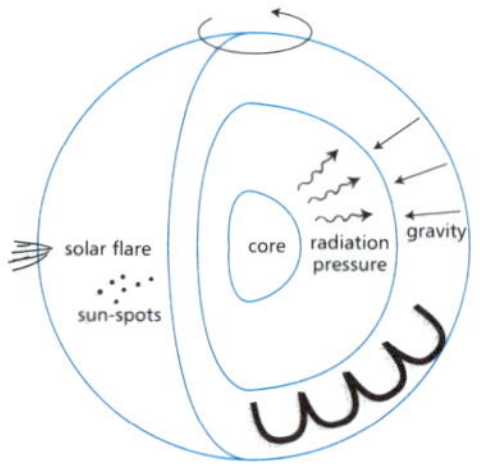

3 A. Contracts to a white dwarf.
  B. Becomes a supernova and then collapses to a neutron star.
  C. Collapses to a black hole.
4

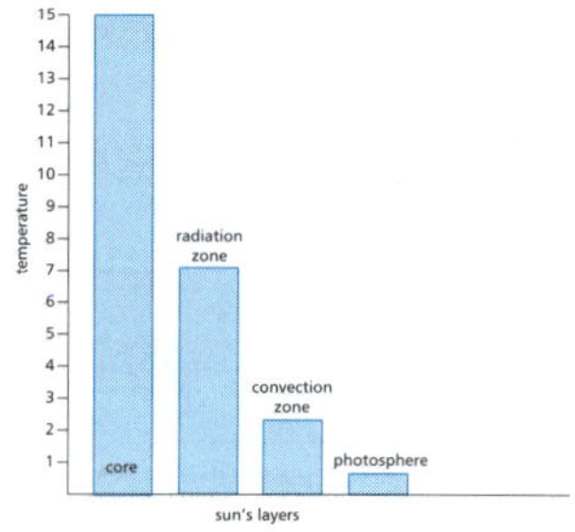

5 a 3 b 2

6

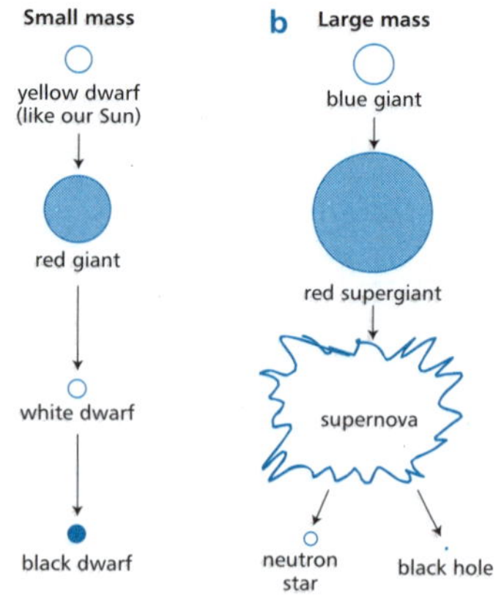

7 b This is the most likely shape for an animal on an ice planet because without any thin areas it would be easiest to keep warm.
8 c
9 A
10 c

Revision questions page 66

1 Sun, light, Sirius, bigger, more
2 a Gravity.
  b Gravity tries to crush the star but the energy from the core makes it expand.
  c It expands, cools, collapses and dies.
  d Elliptical.
3 b
4 b
5 a
6 c
7 a yellow, giant, white, black
  b giant, supergiant, supernova, neutron, black
8

Object	Explanation
Black Hole	Great mass and enormous gravitational pull.
Gravitational Lens	Light rays bend as they pass a massive object.
Quasar	Bright cores of distant galaxies.
Pulsar	Neutron stars that emit radio waves.
Neutron Star	Incredibly dense objects where the electrons have been forced into the nucleus.
Dark Matter	Matter that cannot be seen as it does not give out light.

9 Corona layer
Chromosphere layer
Photosphere layer
Convection Current layer
10 d
11 a no
  b (9 461 × 30 000) trillion km
12 Milky Way, nuclear fusion, nebula, expand
13 a The reaction whereby hydrogen atoms join together to form helium. Each time this happens some mass is lost but large amounts of energy are released.
  b Hydrogen → (Nuclear Fusion) → Helium + Energy
14 a Between Mars and Jupiter.
  b Yes.
  c A comet is made of ice, an asteroid is made of rock. A comet has an elliptical orbit.
15 a 3.5 billion years ago.
  b 4.5 billion years ago.
  c If the Sun has used about half of its mass it should keep shining for another five billion years.

Chapter 7

Science skills page 73

1 D, B, A, C.

2 a Acidic.
  b The gases combine with rain to form a weak acid.

3 a

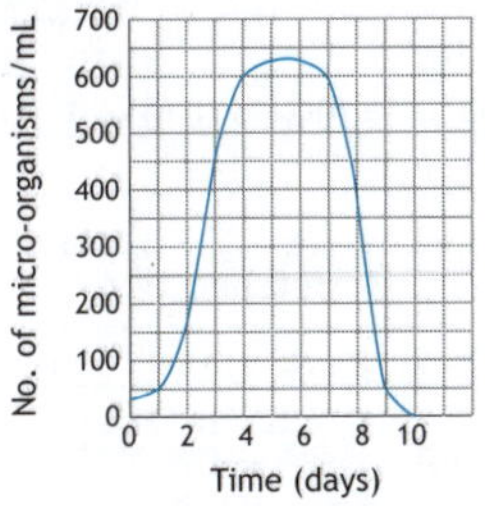

  b Rapid population growth.
  c The population growth has stabilized.
  d Population numbers falling due to competition for resources (food).

4

Pollutant	Source	Release	Effects
CFCs	domestic industrial	aerosol sprays	destroys the ozone layer
sulphur oxides	domestic industrial	burning fossil fuels	acid rain
nitrous oxides	domestic industrial	burning fossil fuels	acid rain
carbon dioxide	domestic industrial	burning fossil fuels	greenhouse effect
heavy metals	mostly domestic	car fumes	accumulation in plants and animals

5 a 1850.
  b The Industrial Revolution.

6 a Beans.
  b Sunflowers.

7 a Over the last 200 years, since the Industrial Revolution.
  b Not long!
  c Because hygiene and medicine improved.

8 a Oil and black coal.
  b Brown coal.
  c Black coal, oil, brown coal and natural gas.
  d Because these fuels are more abundant and easier and cheaper to mine.

9

Resource	Urban	Agricultural	Natural
energy in	much	some	little
energy wastage	much	some	little
materials in	much	some	little
recycling	little	some	much

10

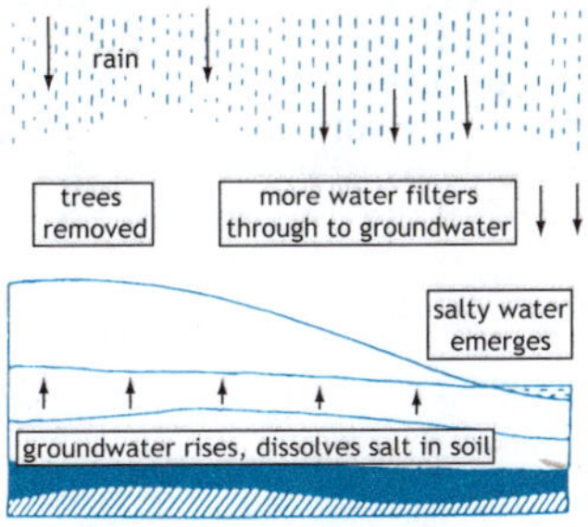

11 Nitrous oxide, CFCs, halons, carbon dioxide and methane.

12 a Because of deforestation and the burning of fossil fuels.
  b Because of the decomposition of organic rubbish and the increase in cattle and sheep numbers.

13 Low nutrient soils erode quickly with no roots to keep the particles together. Rain will wash minerals away and animals will become extinct. Farmers cannot afford fertilisers, cash crops will fail and families suffer malnutrition.

14 a Car exhaust.
  b Peak hour in the city occurs twice a day and involves many cars.
  c Ozone uses energy to build up.

Revision questions page 76

1 Intelligence, tool making, use of fire and teamwork.

2 Carbon dioxide.

3 A change in climate, sea level changes, and animals unable to adapt leading to possible extinction.

4 The enormous demand for fossil fuels since the Industrial Revolution.

5 The clearing of vegetation removes the roots that hold the soil together so the soil washes away. There are no plants to slow the flow of water so flooding occurs.

6 Material wealth and security used to depend on access to the land. In an industrialised world more work, and therefore money, is available to people in cities.

7

Inputs to a city	Outputs of a city
land	organic waste
materials	waste water
water and food	atmospheric pollution
energy	noise pollution

8

Pollutant	Effects
heavy metals	can accumulate in food chains
crude oil	kills sea birds and marine organisms
plastics	kills marine organisms
sewage	over-enrichment of coastal waters — disease risks

9 Drinking water will need to be treated and algae will reduce oxygen levels; loss of important species and algae may restrict water flow and navigation.

10

Primitive man	Controlled by the environment.
Hunter gatherer	Man starts to control the environment.
Agriculturalist	Man has increased control over the environment.
Industrialist	Great control over the environment but with many undesirable consequences.

11 a Grass roots no longer bind the soil and productive land turns to desert.
  b Not enough food; disease and drought conditions.
  c Fence off sections for controlled grazing and use dung for fertilizer.

12 1. An increase in the use of resources and energy.
  2. An increase in the production of waste.
  3. An increase in population growth.
  4. Growth of cities.
  5. Widespread changes to the natural environment.
  6. Western nations build more wealth.

13 nutrients, fertilisers, environment, pollute, salinity.

14 a There are more cars per person now than in the past so we expect this to rise in the future.
  b If more cars are produced then pollution should increase, however, new technology may produce less polluting engines.
  c 0.45 cars per person.

15 a

16 hydrogen

17 a UV can disrupt nucleic acids in living tissue, causing genetic errors and cell malfunction leading to cancers.
  b Chlorofluorocarbons.

18 a It changes into gases that may pollute the atmosphere.
  b The oceans.
  c We must change all our waste into harmless substances and return them to the air, the land or the oceans.

19 a Development that is compatible with the rest of the environment such that the basic needs of all are met.
  b Intergenerational equity, conservation of biodiversity and ecological integrity.

20 a Conservation involves the careful maintenance, management, and preservation of living things and natural resources to make sure that we harm the natural world as little as possible.
  b Reducing pollution and using renewable resources.

Chapter 8

Science skills page 84

1 a wavelength
b amplitude
c crest
d trough
e rarefacti
f rest position
g compression

2 a Blue and red.
b Green and yellow.
c Green.

3 a refraction
b interference
c reflection
d diffraction

4 a 1cm
b 20cm
c 1m/s
d

speed	unit	frequency	unit	wavelength	unit
1	m/s	50	Hz	2	cm
0.5	m/s	10	Hz	5	cm
25	m/s	20	Hz	1.25	m
500	m/s	100	Hz	50	m

5 (B)

6 a, b, c

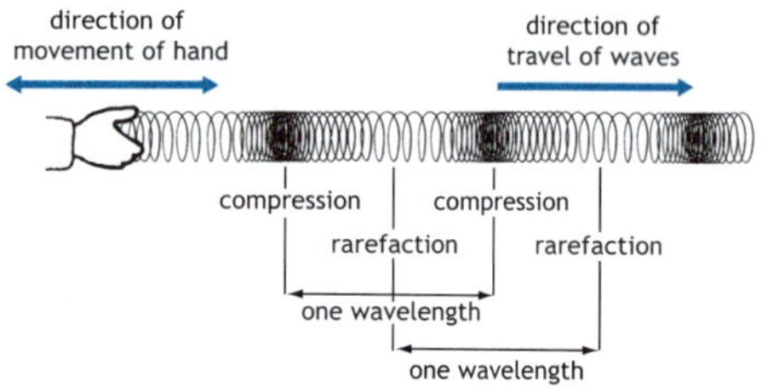

d The distance between two compression or rarefaction zones.
e At B and C.

7 a 1Hz
b 2m
c The knot will move up and down.
d They are reflected.
e Energy is used up in friction and a little heat.

8 c

9 a 2m
b 0.3 m/s
c 15 Hz
d 0.06 m

10 a Gas.
b Solid.
c Because sound waves depend on particle movement. Particles are closest together in solids and furthest apart in gases.

11 a 300,000,000 m/s
b 600m

12 a About 20 — 10,000 Hz.
b About 1000 — 1 000 000 Hz.
c Pigeons and elephants.
d Moths and bats.

13

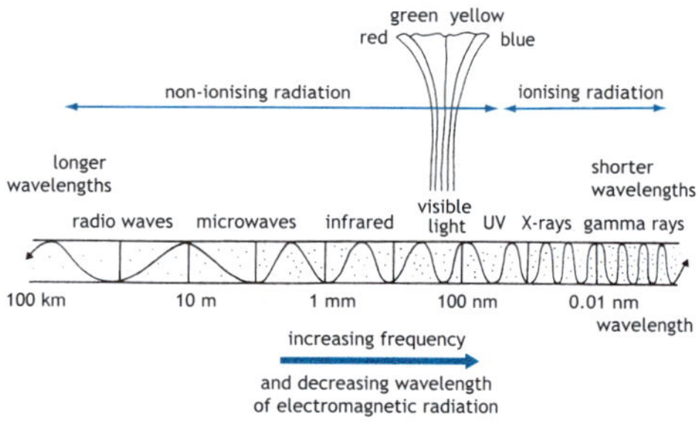

14

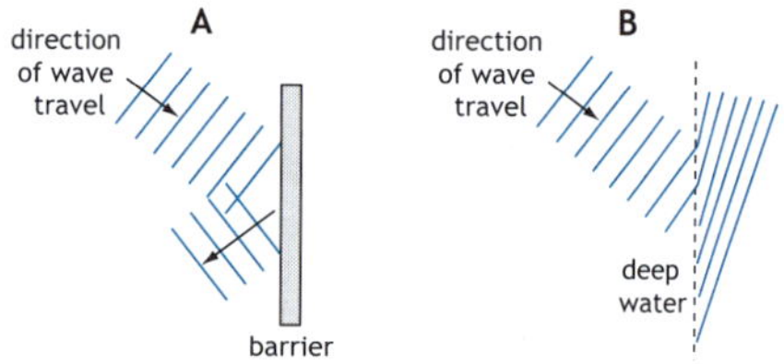

15 a 0.0028m
b 0.0072m
c 66.4m
d 3.0 m

Revision questions page 87

1 a A = longitudinal wave
B = transitional wave
b mechanical

2

Term	Explanation
crest	Top of a wave.
amplitude	Largest distance a particle moves from the surface.
compression	Coils or particles are close together.
trough	Bottom of a wave.
rarefaction	Coils or particles that are stretched out.
wavelength	Distance from one crest or trough to the next.

3 energy, frequency, wavelength, refraction, interference, transverse, longitudinal, longitudinal, light

4 Seeing and hearing.

5 a Depth.
b Sonar.
c Dolphins and bats.

6 whole, back, forth, longitudinal, same.

7

Wave Type	Description
mechanical	Disturb material, or particles.
transverse	The particles oscillate at right angles to the direction the wave is traveling.
electromagnetic	Does not need particles to travel.
longitudinal	The particles oscillate along the direction of wave travel.

8

	Definition	Unit
Speed	The distance travelled by a wave in 1 second.	m/s
Frequency	The number of waves produced per second.	Hz

9 Visible rays — used for seeing, photosynthesis, photography and astronomy.
X-rays — used in radiography, treating cancer, finding faults in metals and examining crystals.
Infra red — used for taking temperature pictures and finding people in collapsed buildings.
Gamma rays — used to measure the thickness metals and sterilising medical equipment.

10 amplitude

11 3.0 m/s

12 compressions, rarefactions, together, apart, frequency

13 A = destructive interference
B = constructive interference

14 a TV waves.
b Long waves (radio).
c Radio reception will be good; TV reception poor.

15 light, frequency, energy, cutting, surgery

16 a A = longitudinal
B = transverse
b They are both mechanical waves.

17 a A, C, E
b B, D, F
c 8
d 1

18 Echoes are sound waves that bounce off surfaces. Humans and animals use echoes to estimate distance.

19 The particles go up and down but the wave moves in a longitudinal direction.

20 4 seconds

Test B page 91

1 a Substances, colour, pH, more, less, equal.
b

Test tube	Colour with universal indicator
C	strong acid
B	medium acid
D	weak acid
A	neutral
E	alkaline

c A = acid B = alkali C = acid
D = neutral E = alkali

2 a chemical, physical, reverse, reactions
b A = chemical
B = physical
C = chemical
D = physical
E = physical

3 a salt and water b alkali c neutral pH

4 a no b yes c physical

5 a no b rusting and respiration
c It carries its own compressed oxygen with it.

6 a Nitric acid and sulphuric acid.
b Because they would react with other substances.

7 a

8 b

9 a red giant, white dwarf
b 4.5 billion years ago
c 3.5 billion years ago

10 a 100 000 light years b 160 000 light years
c Large Megellanic Cloud d 180 000 light years

11 a Nuclear fusion.
b Hydrogen nuclei fuse together and release energy.
c They are slightly cooler areas on the Sun's surface.
d Yellow dwarfs.

12 Big Bang, expand, universe, expanding

13 a The temperatures increased.
b The temperature decreases as the distance increases.
c The temperatures would be much colder.
d Possibly, but not as we know it.
e Water and the atmosphere.

14 a Black holes, neutron stars, pulsars, quasars, dark matter,
b gravitational lenses.
c Neutron Stars.
d Pulsars.

15

Issue	Issue/concern	Solution
energy use	usage is high and increasing	renewable sources
gas emission	increasing — UV reaching the Earth	engines that do not use fossil fuels, remove CFCs and climate changes
waste production and disposal	production of huge amounts of non-biodegradable waste	biodegradable waste and more regulations
greenhouse	the Earth retaining heat	reduce the burning of fossil fuels
ozone	more UV reaching the Earth	reduce CFCs

16 sulphur dioxide, nitrous oxides, sulphuric acid, nitric acid, acid

17 a
- population growth
- high standard of living
- life style in industrialised countries
- gas emissions
- non-biodegradable waste

b
- reduce population growth
- use renewable energy resources
- stringent environmental impact regulations on all industries
- legislation to ensure the production of biodegradable waste

18 a Energy.
b Petrol, oil, coal, diesel and gas.
c Because they were formed millions of years ago from dead animals and vegetation.
d Carbon and hydrogen.
e Carbon dioxide and smoke.
f The greenhouse effect.

19

Pollutant	Source	Effect
sewage	domestic, industrial, agricultural	seriously reduces oxygen levels
fertilisers	agriculture	blooms of blue-green algae
pesticides	agriculture	kills plants and animals
detergent	industrial, domestic	kills valuable decomposers

20 a Better farming and therefore increased food supply.
b Increased hygiene, health care, food production, industry and housing.

21 a The species composition 50 000 years ago was fairly equal. Now rainforest angiosperms dominate.
b This has only been the case for the last 5 000 years.
c In the last 10 000 years rainforest angiosperms have increased and stablised, the Gymnosperms have increased slightly and the sclerophyll forests decreased markedly but are increasing a little now.
d Climate change, particularly rainfall patterns.

22 c

23 a Yes.
b Canada.
c Bangladesh.
d Environmental disaster.
e They would have more money to spend on buying more of your goods and you would make much more money, however, the Earth's resources would be put under more strain producing enough of your good to sell.

24 a The crust.
b Surface waves.
c A transverse wave.
d The mantle and the core.
e Only the mantle.
f Stations A, B, C and D.
g B.

25

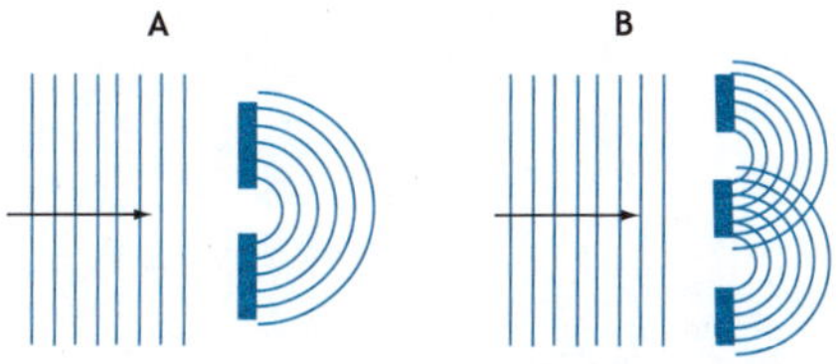

26 a 2cm b 3Hz

27

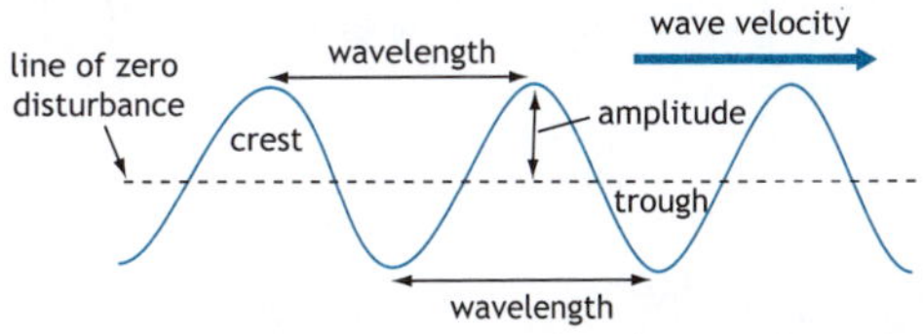

Notes